A Sparrow's Life's as Sweet as Ours

IN PRAISE OF BIRDS AND SEASONS

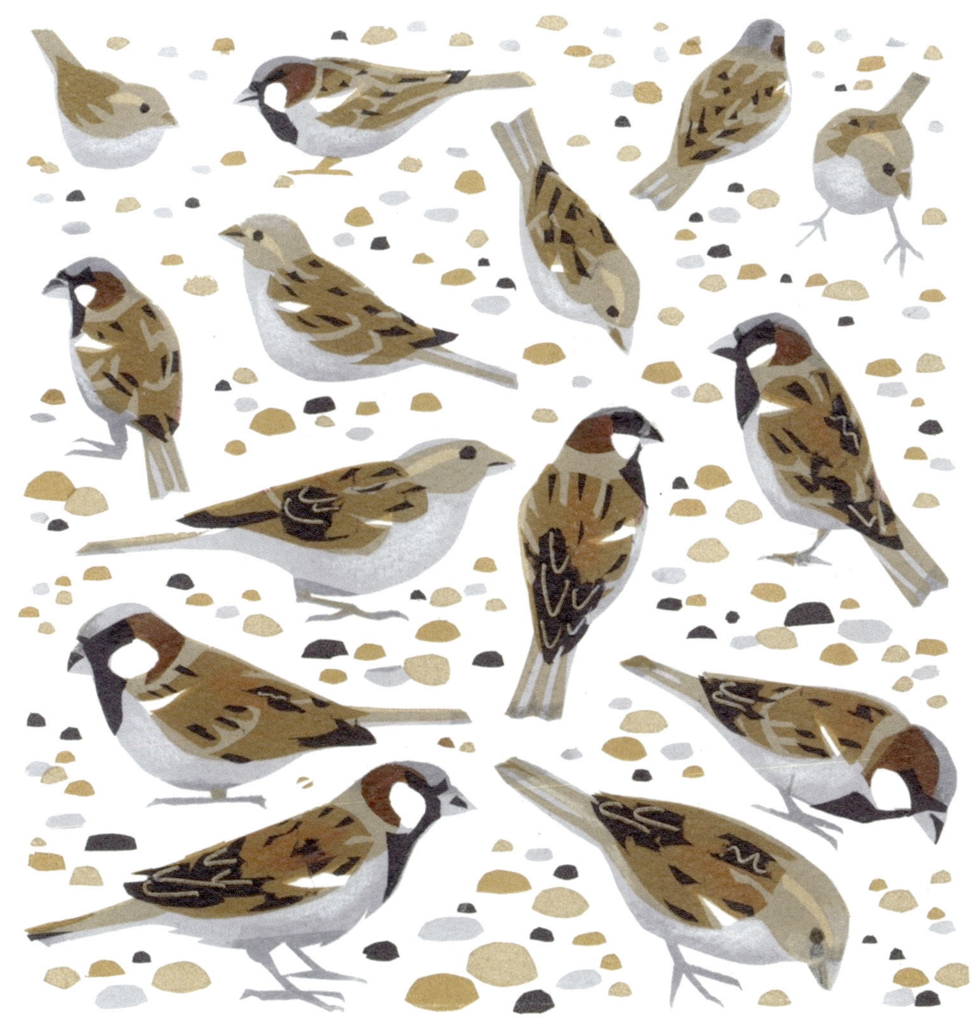

CARRY AKROYD AND JOHN McEWEN

BLOOMSBURY WILDLIFE

LONDON • OXFORD • NEW YORK • NEW DELHI • SYDNEY

To Gordon
who puts up with a lot

To Iris, Innis, Oscar and Mabel
and to the memory of
their devoted Grandma, Jill McEwen

BLOOMSBURY WILDLIFE
Bloomsbury Publishing Plc
50 Bedford Square, London, WC1B 3DP, UK
29 Earlsfort Terrace, Dublin 2, Ireland

BLOOMSBURY, BLOOMSBURY WILDLIFE and the Diana logo are trademarks of
Bloomsbury Publishing Plc

First published in Great Britain 2019

The material in this book has been adapted from *The Oldie* magazine's Bird of the Month columns.
Bloomsbury Publishing would like to thank *The Oldie* for their support of this book.

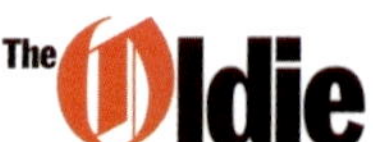

A catalogue record for this book is available from the British Library.

Library of Congress Cataloguing-in-Publication data has been applied for.

ISBN: HB: 978-1-4729-6714-5; ePub: 978-1-4729-6813-5; ePDF: 978-1-4729-6814-2

4 6 8 10 9 7 5 3

Designed by Nimbus Design and Gridlock Design
Printed and bound in India by Replika Press Pvt. Ltd.

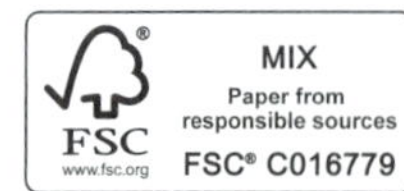

To find out more about our authors and books visit www.bloomsbury.com
and sign up for our newsletters.

JOHN CLARE from 'Summer Evening'

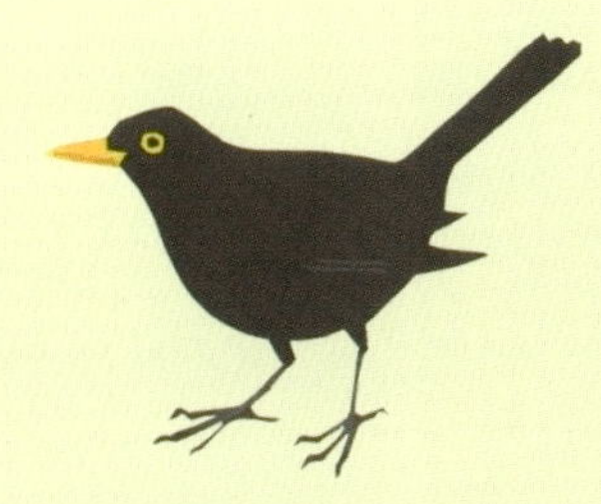

CONTENTS

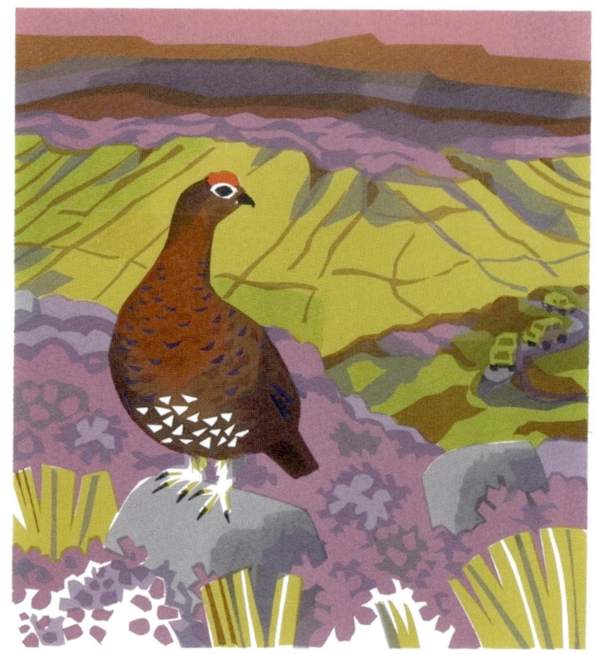

'Gazing out of the window is never a waste of time,' Mark Warman, head of Classics at Harrow School, told his pupils. Two things it shows are the beauty of birds and the similarity of their behaviour to ours.

This book, based on five years of the ongoing Bird of the Month column in *The Oldie* magazine, celebrates 66 British birds representing the four seasons. They are separated by species and habitat. Most are familiar, but the selection is spiced with the odd rarity. Britain has 600 species, either resident or regular migrants – in summer to breed, in winter to feed. So the book is a work in progress.

It is primarily pictorial with complementary text, which includes quotations by poets and writers, from the Old Testament to the present day.

JOHN McEWEN

As a printmaker, screenprinting is a method I enjoy, and it seemed appropriate for this series. I look upon making my monthly bird illustration as a puzzle: I aim to resolve the image in just four layers of colour. (Sometimes it's three or five.) I arrange each layer of colour by using simple cut or torn paper stencils to block areas of the screen mesh before I squeegee ink through onto the image. The bird must be portrayed in an appropriate context, which usually derives from my clearest remembered observation, probably noticing the birds while drawing in the landscape. Sometimes coloured pencils embellish the final image.

After 30 years making work relating to poet John Clare, including linocut illustrations for editions of his poetry, I was invited to be President of the John Clare Society. The title of this book is taken from one of his poems.

CARRY AKROYD
carryakroyd.co.uk

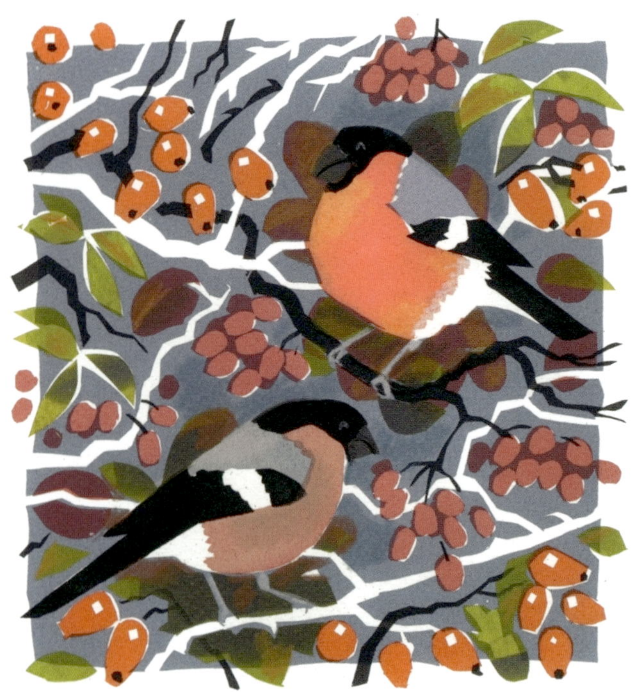

Every four weeks, the Bird of the Month flies into my inbox – one of my favourite moments in *The Oldie* office.

As an amateur birdwatcher, I'm sad that we don't get many birds in our Fitzrovia office (although Harold Pinter's son, Daniel Brand, did once write a letter to us, addressed to the Great Tits of Great Titchfield Street). So, when John McEwen and Carry Akroyd's columns arrive, it's like the whole British bird world descending on central London.

Carry Akroyd's pictures make for a delightful injection of rich colour into *The Oldie*'s pages, not just in the plumage of the birds, but in their closely observed surroundings: the rock crevices favoured by the jackdaw; the woodcock jinking through the snowy woods; the puffins surveying their sea kingdom.

Carry has a unique combination of bird-painting skills. Her pictures are impressionistic – an imaginative leap beyond the straightforward, realist bird guides of my youth. And yet they also capture a bird's jizz – the birdwatcher's term for the overall impression of a bird's size, shape and movement through the air.

Look closely at these pictures and you'll see how Carry captures, in the background, those different flying positions ever so carefully. In her swallows picture, she gets the way they gather on telegraph wires before heading south at the end of the summer; the way they hunch up and gossip away to each other. Then, in the background, she completely nails, in seven images, that most heart-lifting sight, the soaring, swooping and darting, arrow-like movements of the swallow as it scoops up its insect harvest.

We all have our favourite birds – we study them so closely and so regularly that we know their habits better than we do ourselves. Mine is the guillemot; I've been watching them gather at Stack Rocks, near my parents' Pembrokeshire cottage, every summer for 35 years. Carry has got their subtle little movements down to a T: the way they nod, raise their heads, slightly sniffily, and bump right up against each other on their crammed nesting sites.

John McEwen is an expert ornithologist, but he writes

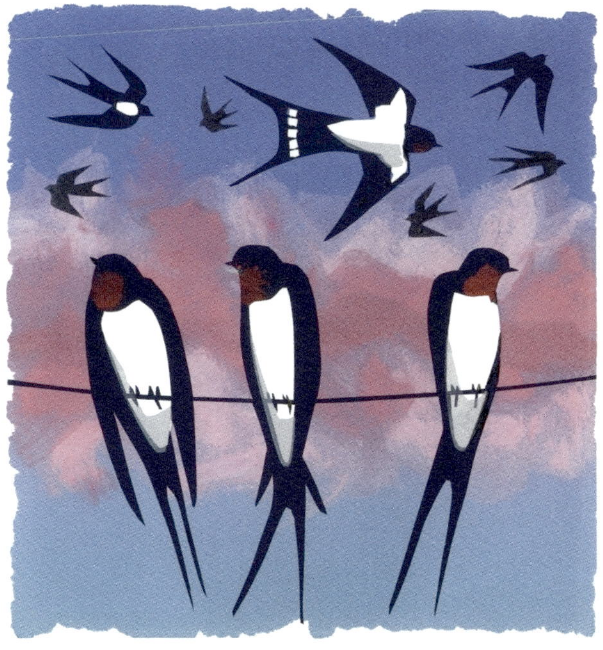

in a way that transcends the dry, verbose, scientific prose of ornithology. A distinguished art critic, he has the aesthete's eye for the movement and plumage of a bird. He sees, quite literally, the poetry of birds: quoting from the best poets, who are so often bird lovers, finely attuned to birds' calls; whether it's Norman MacCaig on the curlew ('Music as desolate, as beautiful, as your loved places') or John Clare on the nuthatch ('In summer showers a skreeking noise is heard').

Like all the best journalists, McEwen empathises with his readers – and, in this case, their everyday experiences with birds today: whether it's with the mistle thrush who made its untidy nest in a traffic light in Beeston, Leeds, in 2010, or the pied wagtails who roost in Tesco car parks, marinas and petrol stations.

All British bird life is here.

HARRY MOUNT Editor of *The Oldie* magazine

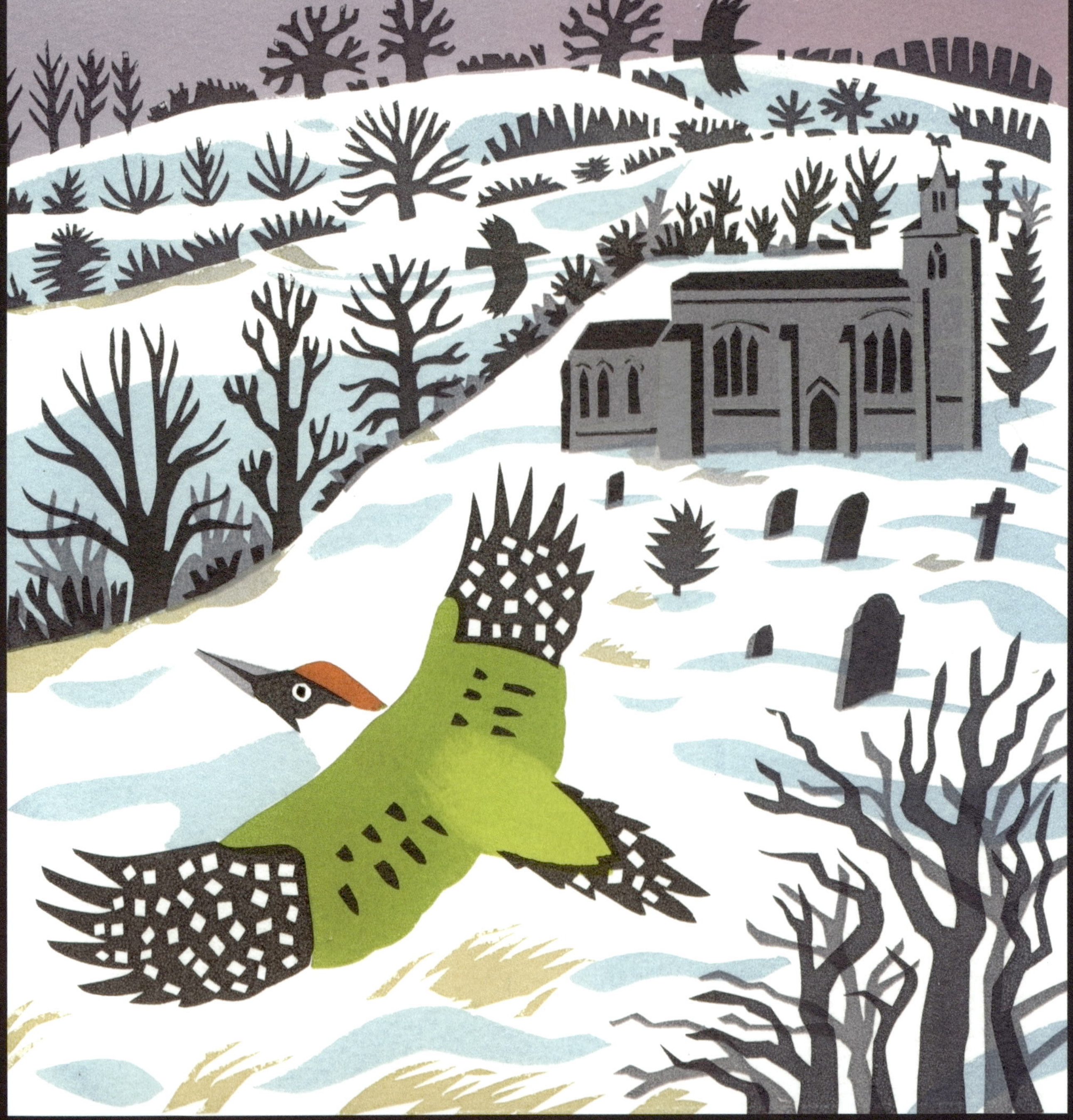

THE GREEN WOODPECKER

Suddenly, like an arrow from the East,
Smart as new paint appeared our gorgeous guest,
Green in his plumage, scarlet on his crown,
Strolling about a tree-trunk upside down,
Tat-tapping busily with beak on bark,
His mind imagineless, his purpose dark
To us who watched him, startled out of speech,
Exchanging secret glances each with each.
So all that day, being young and fearing not
Lest custom dull the treasure we had got,
All that day long we jingled, as we went,
New-minted coins of bright astonishment.

GERALD BULLETT, 'Woodpecker'

The green woodpecker (*Picus viridis*) can be difficult to spot in rough pasture, but there could be no more festive visitor to a Christmas lawn. Its yellow rump can even persuade optimists it is a golden oriole, sadly no longer a British summer migrant. Sexes are similarly plumaged except that the female lacks the red 'moustache'. Nor could there be anything more welcoming on a Christmas Day constitutional than its piping 'kew-kew-kew'. Tennyson caught the resemblance to human cackling:

Her rapid laughters wild and shrill,
As laughters of the woodpecker
From the bosom of a hill.

From 'Kate'

'Yaffle', its colloquial name, expresses the spirit not the sound of the penetrating pipe – uttered in flight as it bounces through the air in the woodpecker way. Hecco, hewel, popinjay (more often a parrot in heraldry), wodewale are some of its old names. Carry Akroyd's bird flies past St Margaret's, Luddington in the Brook (also see The Goldfinch, page 38).

In folklore, woodpeckers in general – the green in particular – are linked with water, hence one of its many old names 'rain bird'. For farmers who cleared the wilderness, the green's appearance on newly tilled ground must have seemed a blessing, with rain its consequence. Woodpeckers are marooned survivors of the 6,000 BC tsunami, which cut off England from continental Europe. The sea still excludes them from Ireland.

In Britain, greens are concentrated in south-east England, where they continue to multiply, dwindling west and north and ceasing altogether beyond the Highland line. The first record of a breeding pair in Scotland was 1951.

They nest in tree holes dug noisily by themselves, usually in apparently healthy but rotten trunks. Unlike the great spotted woodpecker, males do not drum their beaks against wood as a mating call. Courtship display involves the raising of crimson crests and fanning of domino wings and tails as the birds play hide and seek.

Trees offer plentiful grubs, but ants are its preferred food, making lawns, urban parks and pasture its favourite feeding grounds. Its choice as the logo for Bulmers Woodpecker Cider stems from England's traditional apple orchards being a favourite habitat. The ants adhere to its sticky, four-inch-long (10cm) tongue, which, when withdrawn, coils inside the skull like a spring. Records show one bird dug 2½ feet (76cm) to reach an ant nest, and a brood of seven chicks consumed 1½ million ants and their pupae. Pairs are monogamous. Each takes responsibility for half the brood and is solitary outside the breeding season. Frozen winter ground can lead to fatal starvation.

In 2015 Martin Le-May's photograph of a green flying off with a weasel on its back proved a news sensation. It revealed the danger of the bird's ground feeding. Happily, the weasel was dumped when its bearer crash-landed.

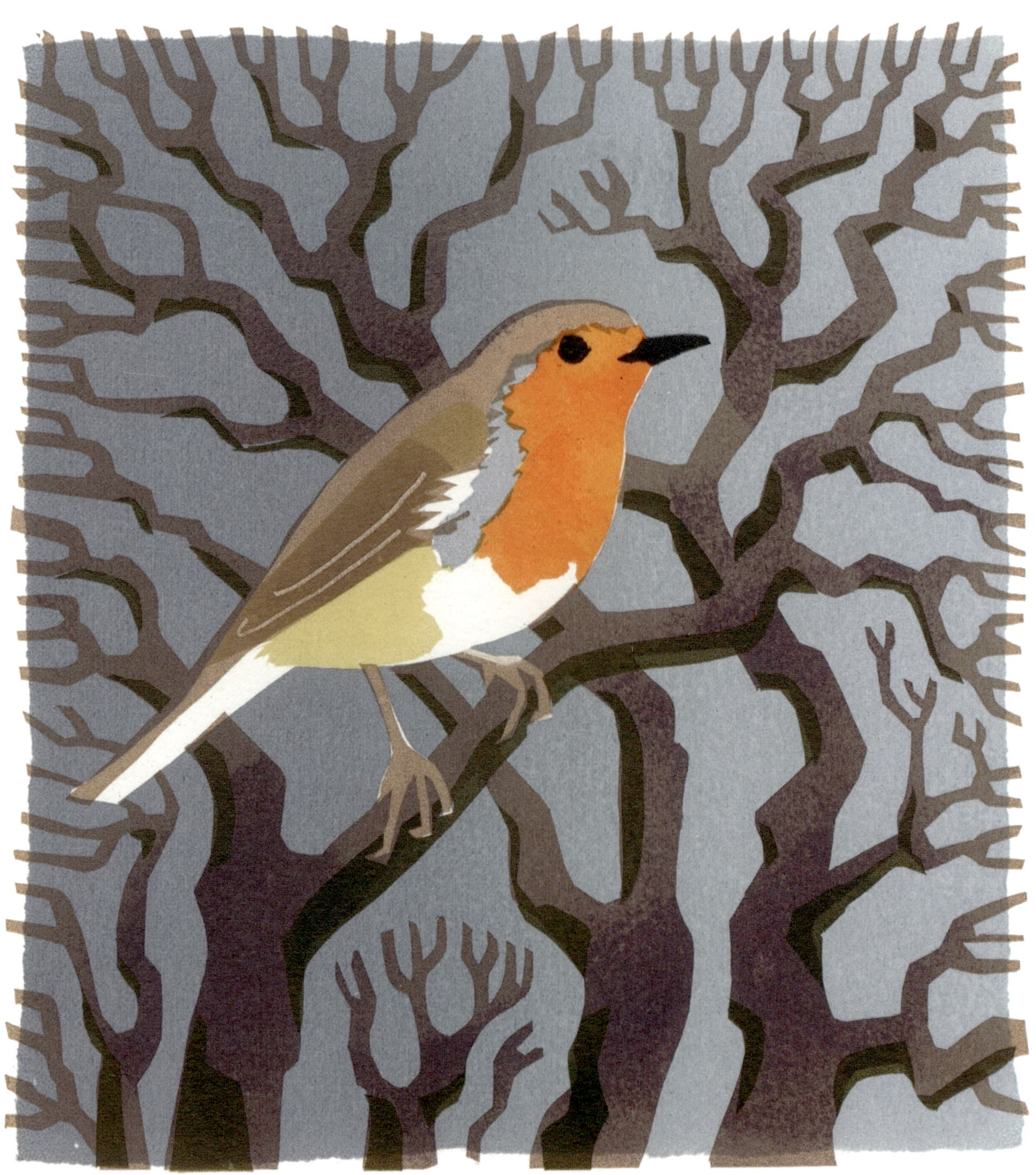

THE ROBIN

In 2015 the robin (*Erithacus rubecula*), reclassified a flycatcher (Muscicapidae), was popularly reconfirmed as the UK's official national bird. The vote was organised by 'urban birder' David Lindo: 'Despite being a seemingly friendly bird, the robin is…very defensive of its territory and I presume that reflects us as an island nation that we will stand our ground.' Those emblematic Christmas cards surely helped it top the poll, in which case it was a very suitable choice because it has long had sacred associations:

> *The robin redbreast and the wren*
> *Are God almighty's cock and hen*
> ANON, traditional

St Serf of Culross had a pet robin, that perched on his shoulder as he prayed. His favourite pupil was St Mungo (d. 614), founder and patron saint of Glasgow. Jealous classmates killed their teacher's bird and blamed it on Mungo, who miraculously restored it to life. The resurrected robin is one of his four miracles that constitute the city's coat of arms.

In Wales, robins are considered birds of ill omen, perhaps because they are often seen around churches, not least when grave-diggers are at work. Walter Scott hints at the mortal connection in 'Proud Maisie':

> *Proud Maisie is in the wood,*
> *Walking so early;*
> *Sweet robin sits on the bush,*
> *Singing so rarely.*
> *'Tell me, thou bonny bird,*
> *When shall I marry me?' —*
> *'When six braw gentlemen*
> *Kirkward shall carry ye.'*
> From the novel *The Heart of Midlothian*

Robins nest everywhere, even completing one between breakfast and lunch in the pocket of a gardener's discarded coat (Mark Cocker, *Birds Britannica*). Its sacred reputation and familiarity – the 'curious' robin known to all gardeners, bonfire-makers and foresters – have afforded it some protection as a good omen.

> *A Robin Red breast in a Cage*
> *Puts all Heaven in a Rage.*
> WILLIAM BLAKE, from 'Auguries of Innocence'

Ruddock, its ancient name, used by Chaucer, was displaced by redbreast (fifteenth century) and robin (sixteenth century, from the Frisian *robyn*). The first postmen were called robins after their red coats, an association which in the 1860s inspired the first Christmas cards. The pleasure-giving robin bringing Christmas good cheer – it was usually shown carrying a letter in its beak – was enhanced by its approachability. Lord Grey of Fallodon (1864–1933), landowning fly fisherman, author, bird lover and, as Sir Edward Grey, Foreign Secretary at the outset of the First World War, wrote in his classic book, *The Charm of Birds*, 'Any robin can be tamed.' and provided instructions.

The robin's song is its special beauty. We are told it is purely territorial, yet it matches the seasons: joyful chorister in spring and summer; 'wildly tender' (Emily Bronte) soloist in autumn and winter. W. H. Davies wrote in its praise:

> *Robin on a leafless bough,*
> *Lord in Heaven, how he sings!*
> *Now cold Winter's cruel wind*
> *Makes playmates of poor, withered things.*
> *…*
> *If these crumbs of bread were pearls,*
> *And I had no bread at home,*
> *He should have them for that song;*
> *Pretty Robin Redbreast, Come.*
> From 'Robin Redbreast'

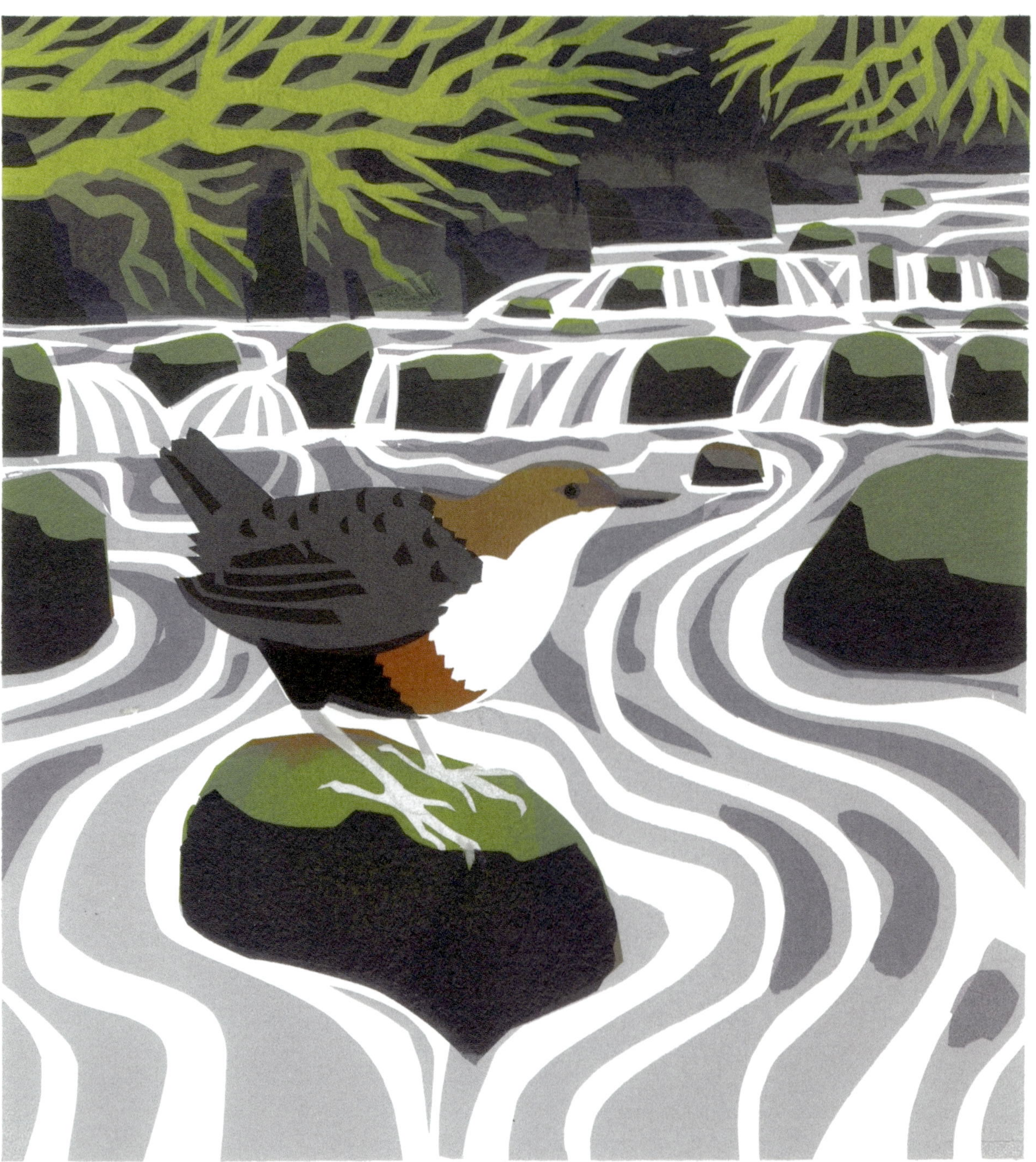

THE DIPPER

The dipper (*Cinclus cinclus*) was formerly called a water ouzel, although it is no relation of the ring ouzel (a thrush). Among birds, it is one of the most unusual: the only fully aquatic songbird, in that it actually feeds and can escape underwater – although not always: Gareth Thomas of Ludlow reported one fished out by a lesser black-backed gull (*British Birds*, Feb 2018). Among other distinctions, it sings at every season, not least the depths of winter, and its first eggs are often laid in February.

It was winter, near freezing,
I'd walked through a forest of firs
when I saw issue out of the waterfall
a solitary bird.

It lit on a damp rock,
and, as water swept stupidly on,
wrung from its own throat
supple, undammable song.

It isn't mine to give.
I can't coax this bird to my hand
that knows the depth of the river
yet sings of it on land.

KATHLEEN JAMIE, 'The Dipper'

Film reveals how the dipper can swim to feed, using its wings like flippers. Science, how uniquely it is adapted to water. Its dense plumage is waterproofed by an enlarged preen gland. Its bones are not hollow like other perching birds', but solid to reduce buoyancy. Membranes protect its eyes, and special muscles enhance underwater focus. Flaps seal its nostrils, and haemoglobin concentration in the blood stores oxygen; the rate its body at rest burns energy is one-third slower than equivalent birds. Glacial water is not a problem. Longer legs and sharper claws enable it to withstand currents. Underwater it forages for larvae, fish eggs and small molluscs for up to half a minute at a time. Even its calls are pitched to be heard above the torrent rush.

The cleaner the water the better, so it favours fast, boulder-strewn, invariably upland streams and rivers. Territories, fiercely defended, can be from a half to two miles (0.8–3.2km) long; the more plentiful the food, the more songs it sings – as few as five and as many as nine. The mossy domed nest is often under a bridge or curtained by a waterfall.

Dippers have cocked tails like their wren relative, and both sexes are boldly plumaged:

The water-woosel next, all over black as jet,
With various colours, black, green, blue, red, russet, white,
Do yield the gazing eye as variable delight
As do those sundry fowls whose several plumes they be.

MICHAEL DRAYTON, from *Poly-Olbion*

The avian Samuel Johnson, in portliness at
least, if too dapper in its russet and white and
slategrey among light's silvers on the wet half-mossed
boulder, midstream, closing the hinges of its legs in
frequent semi-curtseys, the nictitating membrane
flicking white across its eye. Bird of perpetual
motion, among the musics of water. Breastpuffed
conductor, spivved up, taking its constant encores.
Then down it goes again, under, where? – then up,
unexpectedly – there! – on another boulder: a flick
of rotundity, a minor shower of shattered crystal,
then the curtsey, curtsey, curtsey. Thank you, thank
you. You are most kind. *Sometimes a scrap of water*
song, a mingling with the Lugton's trickletones,
rippled away downstream. Then suddenly! – sip! sip!
– goes birring past you upstream, magnetised to the
narrows of clear meandering life among the deserts of
anemone and celandine.

GERRY CAMBRIDGE, 'Cinclus cinclus'

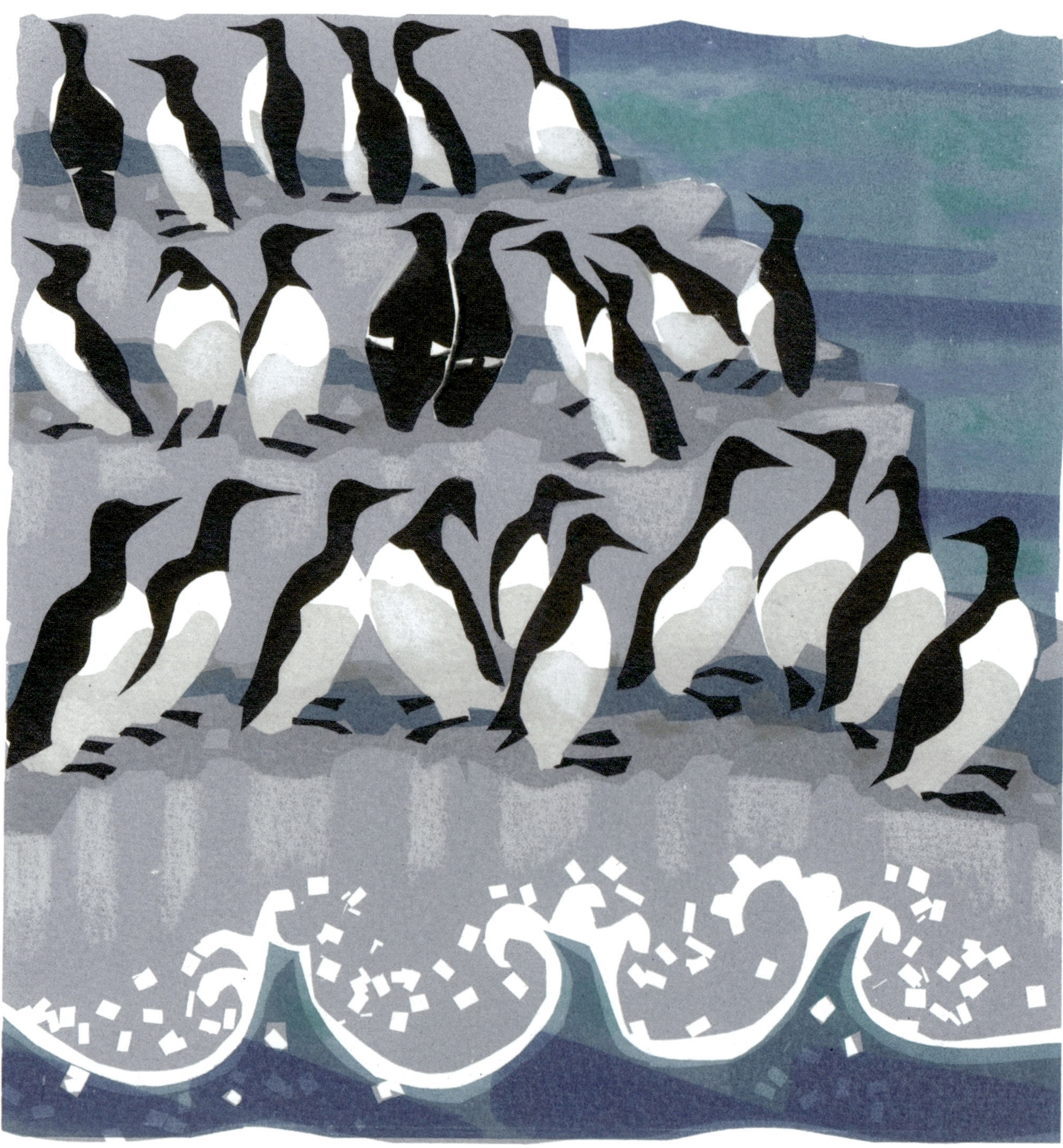

THE GUILLEMOT

The penguin-like auk, the common guillemot (*Uria aalage*) is a summer migrant, so how can it be a Christmas bird? Answer: Ronald Lockley. He established Britain's First Bird Observatory on Skokholm, off Pembrokeshire, in 1933 and described in *The Island* how the guillemots would appear on sunny days in late December, months before their seasonal arrival in March. They would dance in the waves and flutter up to their nest sites. The Pembrokeshire locals, who called them 'eligugs', assumed they were celebrating Christmas.

By then the adults were already in their dark chocolate and white breeding finery – the close-knit feathers tight as a pelt – but there was no mating. Performance over, they disappeared. There had been nothing sexual, just pure joy – a mid-winter clan gathering after the months of moult and ocean wandering; the visit to ledges, an apparent reclamation of that small section where they would lay the single egg and raise the chick in the coming year.

The egg is another guillemot marvel. In his book *The Most Perfect Thing: Inside (and Outside) a Bird's Egg*, Tim Birkhead, who has 'lived and breathed guillemots for 40 years', singles out the egg as extraordinary for the variation of its 'size, colour and pattern'; no two alike. The egg and nestless breeding colonies – the densest of any cliff-dwelling bird – made the guillemots particular victims until modern bird-protection legislation. Across the northern hemisphere, hundreds of thousands of the eggs were taken to make soap, to eat or as art objects. At Bempton in Yorkshire there was an industry of 'climmers', who risked their lives to gather them. The birds were also shot like driven pheasants by boating parties and, for many coastal communities, the meat was a valuable food.

Between the wars, the oologist George Lupton formed a collection of a thousand guillemot eggs. In the wild, smeared faeces obscure their beauty, although its fellow auk, the razorbill, manages to keep its eggs clean. That the egg is pointed to prevent it rolling is a myth. Despite the birds'

nesting shoulder to shoulder, Lockley was amazed a female recognised a returning mate at half a mile.

The guillemot, as Adam Nicolson wrote in *The Seabird's Cry*, a paean to the seabirds of his Shiant Isles in the Minch, is 'super-conservative', annually laying its egg on the exact spot on the same narrow ledge. He has dived with them: 'They come past you in spurts, each wing-thrust jerking them forward'. Capable of 4½ minutes underwater, they can reach depths of 600 feet (183m).

'Guillemots are our barometers of marine well being', Birkhead has written. Often the flightless victims of oil spills, they register every environmental change. Welsh government funding for Birkhead's 25-year research programme on Skomer's guillemots, Skokholm's neighbouring island, ceased in 2013, and in 2014 a storm devastated the colony. An article in *Nature* saved the project. In 2016 numbers on Skokholm were the highest since 1927.

In the first weeks of 2019, 20,000 guillemots were washed up dead along the 186 mile (300km) Dutch coast. The birds were drenched in oil and autopsies revealed no plastic in their stomachs. The container ship MSC Zoe had shed 345 containers off the coast in a New Year storm, but no scientific connection could be made between the spillage and the birds' deaths. A number of stranded starving birds were saved. Similar unexplained guillemot decimations occurred in Holland in the 1980s and 1990s.

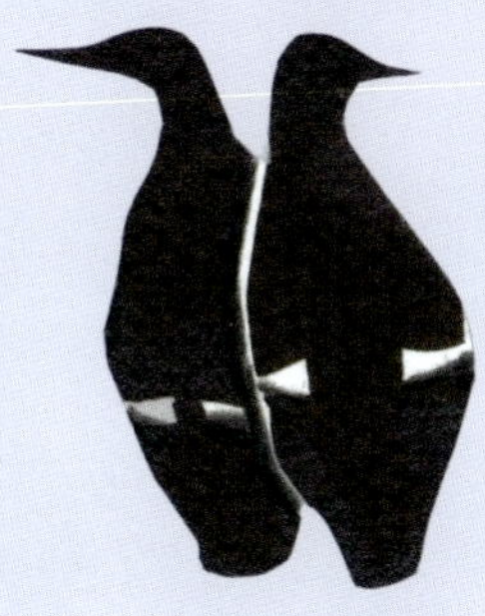

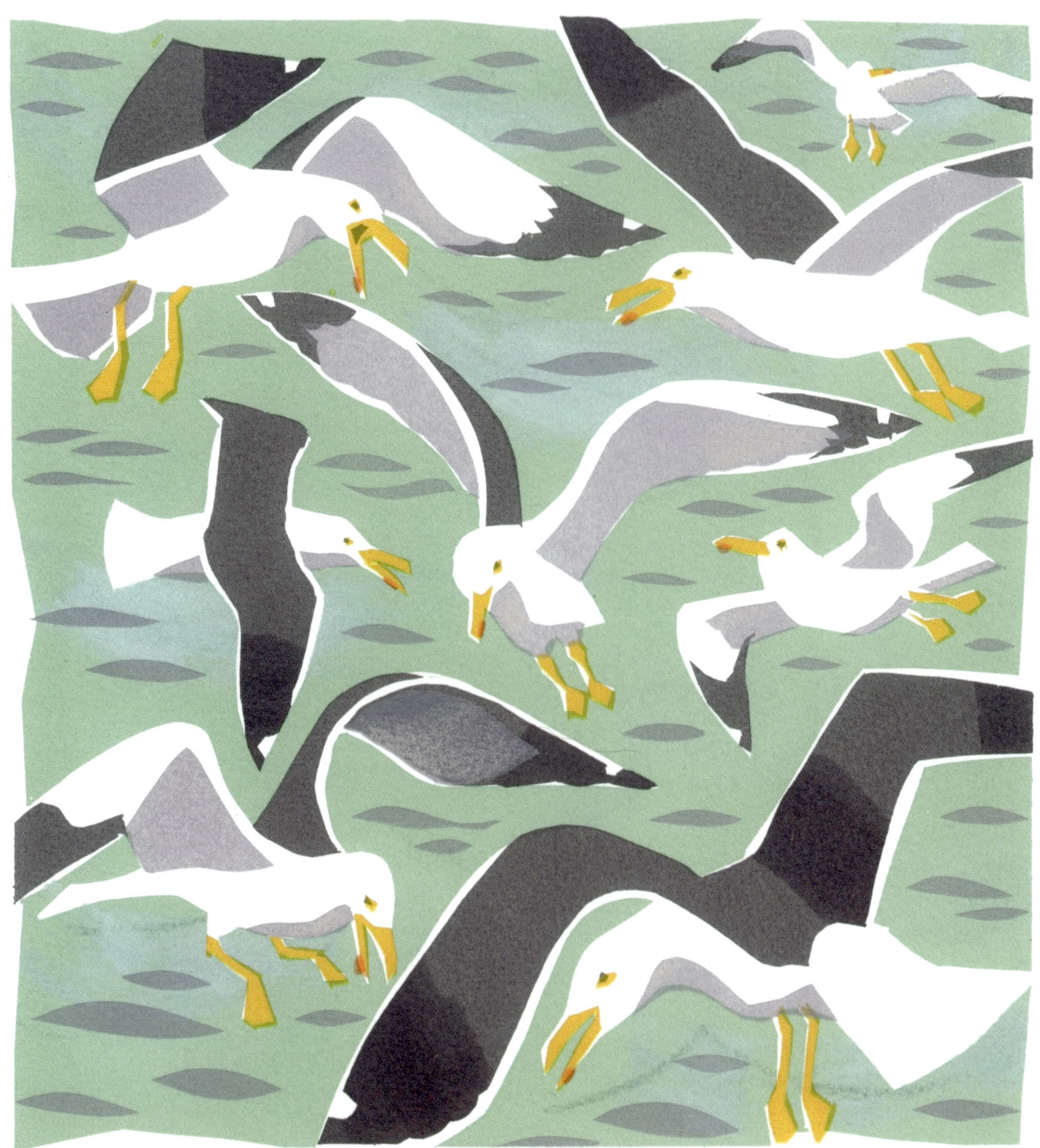

THE LESSER BLACK-BACKED GULL

(gray haddock, falseface skate, flounders
with wrong eyes — they slide into the shallow boxes
with a slip, with a slither, watched by me
and by herring gulls and blackbacks that stroll
the fringe of the crowd like policemen,
like pickpockets).

NORMAN MacCAIG, from 'Centre of Centres'

For 'blackbacks' read the lesser black-backed gull (*Larus fuscus*), a slightly smaller herring gull, with which it is often seen, but with slate-grey mantle and wings. Impossible to confuse with its namesake, the greater black-backed, mightiest of gulls with jet-black in place of slate.

Some lesser black-backeds, like other gulls, have been forced through human interaction to forsake their immemorial home on coast and sea. Norman MacCaig describes them eyeing freshly caught fish. Twenty-first-century high-tech fishing is not so bountiful.

The watershed was the 1953 Clean Air Act, which illegalised waste burning. A bonanza of rubbish dumps and landfill sites, fuelled by the arrival of consumerism, was the result. Recycling and environmentalism in the twenty-first century are turning these into 'country parks'. Gulls have been forced to take to the streets to gulp our trash and litter, bags and all. We derisorily call these always elegant birds 'bin chickens' and 'shite hawks'.

A fair proportion of both species have consequently set up house and become roof-nesters. Herrings have always preferred cliffs; lesser black-backeds sand-duned or flat. Carry Akroyd, sorely harassed by the anxious birds, even had to avoid their eggs along a path on Samson, Isles of Scilly. In town, herrings like something to wedge against like a chimney stack; and lesser black-backeds opt for flat roofs. Urban warmth can encourage nesting from January.

Bristol has been notably popular. Both gulls colonised the islands of Flat Holm and Steep Holm in the Bristol Channel; the lesser black-backeds as summer migrants from the western Mediterranean and West Africa, arriving in April and leaving in September. As James Fisher wrote in 1947, it was 'a summer migrant'. When the islands reached capacity, the surplus settled in the city.

Peter Rock, a Bristolian, was the first person to study urban gulls. In the 1980s there were 100 pairs of Bristol roof-nesters. In 2018 there were 2,500. The same was true nationwide. The last official census in 2000 estimated 31,000 pairs of urban gulls. Twenty years on, Rock reckons these have multiplied to more than 100,000. Yet overall, herring gulls are in red-listed decline, with lesser black-backeds listed as 'No Concern'. Urban lesser black-backeds no longer migrate. Rock's oldest bird is a lesser black-backed. He ringed it as a five-week-old chick and sees it regularly around the city. It will be 30 in 2020. Large gulls take four anonymous years in mud-spattered-looking plumage to attain the beauty of adulthood, a challenge welcomed by ID-fixated 'gullers'.

We are to blame if gulls now encroach to our discomfort. Peter Rock wears a 'seagull jacket' on his rounds, 'so when they fly over me they see I've already been done, so they'll crap over you instead'. But he is devoted to his gulls. As Tim Dee, also a Bristolian, wrote in *Landfill*, 'the meeting of gulls and people is exuberant. It has an excess that could be called joyous'. Except if you disturb a nesting colony.

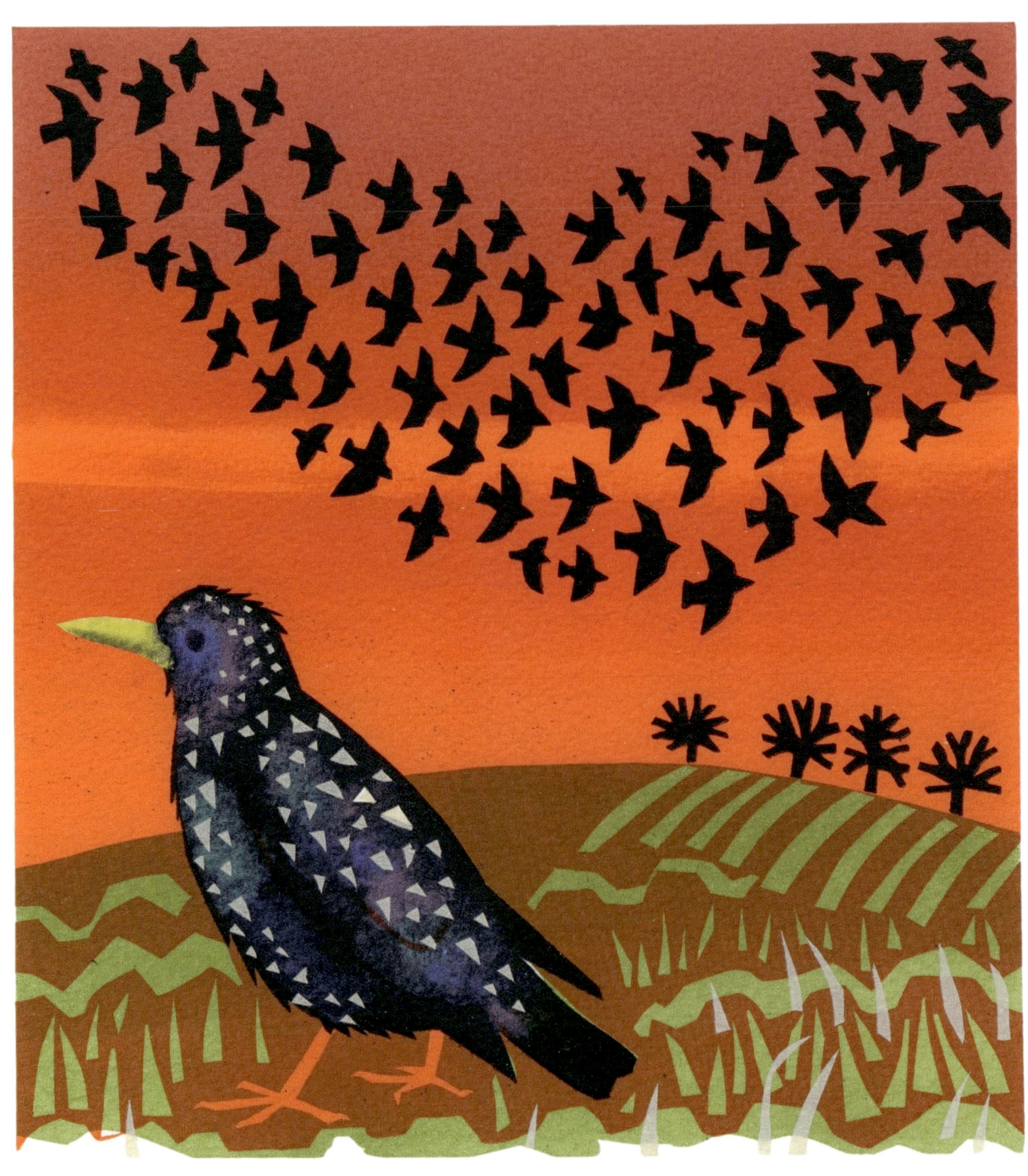

THE STARLING

Winter is the time to see the starling (*Sturnus vulgaris*). The flocks, diminished in the breeding season when beaks turn yellow, have reassembled, boosted by continental arrivals and dun-plumaged juveniles. The Dutch biologist A. C. Perdeck conducted some experiments into bird migration using starlings and chaffinches 'on a larger scale than any before or since' according to the English biologist Rupert Sheldrake (*Dogs That Know When Their Owners Are Coming Home*). In the 1950s Perdeck captured starlings in Holland on their migration from the Baltic to England and northern France and released them in Switzerland, juveniles separated from adults. Normally starlings flock regardless of age. The juveniles flew south-west but parallel to their original route, arriving in southern France and Spain. The adults found their way to England and northern France. As Sheldrake wrote: 'In other words, the adults were showing navigational behaviour similar to that of homing pigeons, dependent on the connection they had made to their winter homes, their "site-imprinting".' The juveniles returned to the Baltic and re-migrated to southern France and Spain. A new migratory cycle was thus established. For Sheldrake, it was further proof of 'morphic resonance' (see The Blackcap, page 55).

Their winter aerobatics defy scientists and inspire poets. In Ned Denny's forthcoming verse interpretation of Dante's *Divina Commedia* he translates the starling reference in *Inferno*, canto v:

As a blue flight of birds on glassy porcelain
is tousled and torn at by an eternal wind,
so they hurtled [...] here, there and everywhere.

From B (after Dante)

Then there is the starling's talent for mimicry.

The resident starling echoed the Capstan double-strength smoker's cough of J. N. F. McEwen (1924–71), the bird artist, down his Ayrshire chimney. Starlings, 'the poor man's parrot', were once popular cage birds. The most famous was Mozart's. He marked its death with full funereal honours. Veiled mourners processed, hymns were sung, and he delivered a graveside elegy, which began:

Hier ruht ein lieber Narr,
Ein Vogel staar.
(A little fool lies here,
Whom I held dear —)

Were the illogical progression and crashing final discord of *Divertimento for two horns and string quartet, K. 522*, his first composition after the bird's death, a parody of its random mimicry? He left no explanation.

Starlings are happy in town and country. In 1890 a pharmaceutical millionaire, Eugene Schiefflin, sought to introduce all birds mentioned by Shakespeare to the USA. Sixty starlings released in New York have resulted in a nationwide, 2-billion strong, pest.

Town centres at sunset used to seethe and bristle with starlings. For the 1963 Leicester-Square premiere of his film *The Birds*, Alfred Hitchcock arranged for a bang to go off, letting rip the roosting starling horde. In 1949 a flock slowed one of Big Ben's 15½ stone (98.4kg) minute hands. In 2019 they are mostly reed-bed roosters. They remain the UK's second commonest garden bird but by 2019 had nonetheless declined 84 per cent in 40 years. Lack of pasture and eating sewage-works earthworms seem two causes of decline. The worms transmit Prozac, which creates lost appetite and libido in the birds, as confirmed by Dr Kathryn Arnold's research at York University. Yet surely this boisterous bird will prevail.

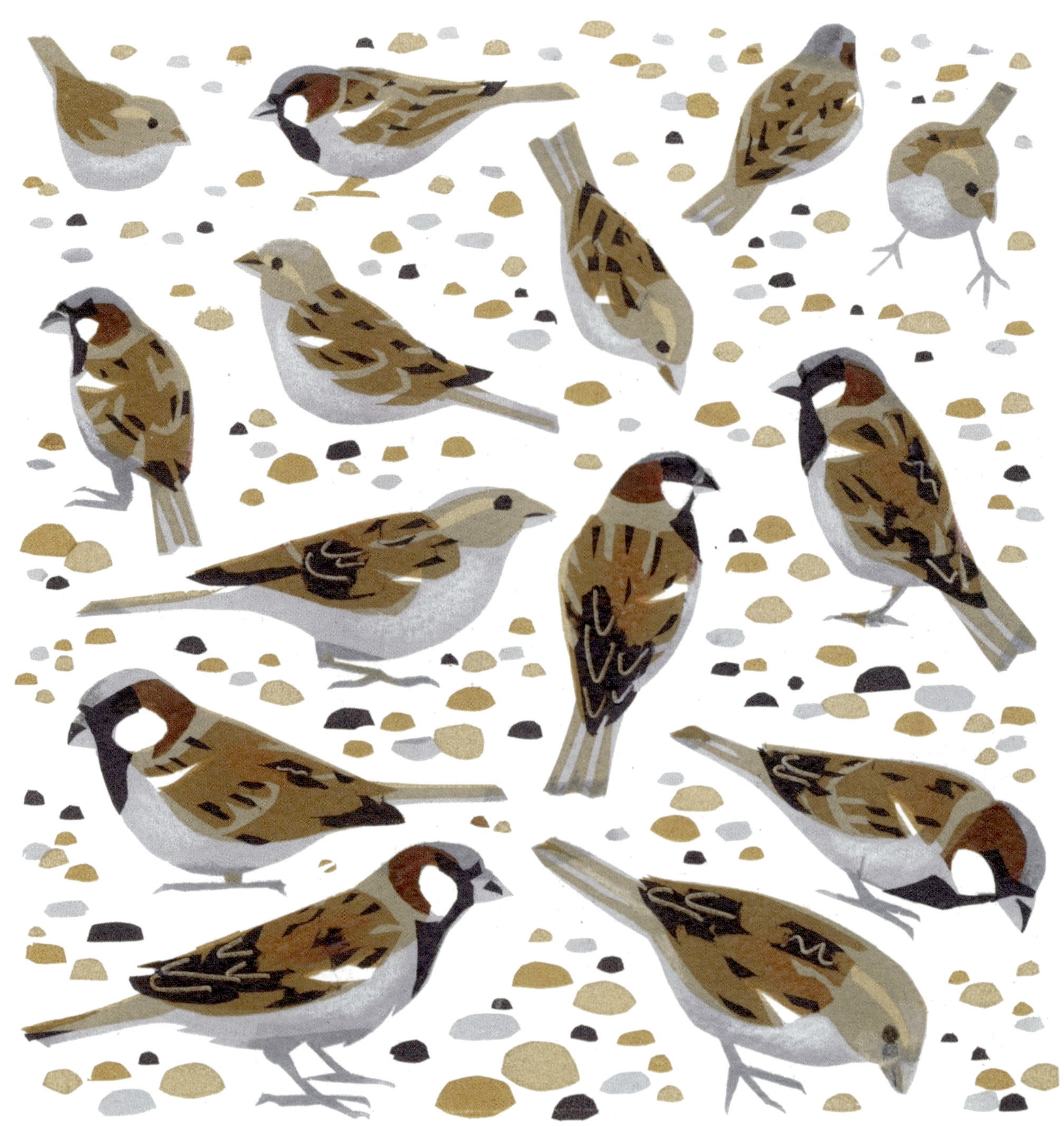

THE HOUSE SPARROW

NORMAN MacCAIG, from 'Sparrow'

No bird divides the experience of old from young as dramatically as the house sparrow (*Passer domesticus*). There was a bounty of a halfpenny a bird (approx 8p, 2017) in the Second World War, so damaging to food supplies were its numbers considered. Their rampant breeding made them for Aristotle 'of all birds the most wanton', also reflected in the 1940 music-hall song 'Wot'cher Me Old Cock Sparrer'. Hence their traditional reputation as an aphrodisiac, both their eggs and meat. Now, especially in town, house sparrow numbers have plummeted. In Paris too, the population has fallen by 200,000 in 17 years despite the bird being so emblematic of the city that Edith Piaf (slang for 'sparrow') adopted its name.

The decline was famously charted by Max Nicholson, founder of today's Worldwide Fund for Nature, who, as a 21-year-old, in 1925 enlisted his brother Basil to help make a complete avian census of Kensington Gardens. They devised 19 sections and on 2 and 4 November spent nine hours counting. The method had been successfully pioneered in America, but never before tried in Britain. It revealed 30 species and 3,982 birds, of which 2,603 were house sparrows. A month later they did it again. An eight-sparrow difference proved the method's accuracy. The next census, in 1948, revealed 885 sparrows; in 1975, 544. Decline became a crash in the 1990s.

In 2000, attended by the now-legendary 96-year-old Nicholson, there were eight sparrows. It was a sunny November morning and most of the deciduous trees were bare. At the first census, they had been thick with leaf. 'Global warming!' was Nicholson's emphatic conclusion for the birds' decline, combined with his hunch that they had a suicidal tendency to celibacy once numbers fell below a certain level. Sparrows duly disappeared from central London north of the Thames. Then in 2015 three were sighted in Kensington Gardens. Revival was as inexplicable as decline. In my Camden neighbourhood the local colony had disappeared by 2010; since 2015 it has risen to raucous double figures. Yet in Honor Oak and other south London boroughs that merry 'chirrup' ('Cheer up! Cheer up!') greeted one throughout north London's sparrow-less years. Elsewhere in Britain decline was less marked. The sparrow topped the 2018 UK Big Garden Birdwatch, when 420,489 British bird lovers took stock in the last weekend of January (see The Blue Tit, page 25).

The oarsman Alex Gregory, double Olympic champion in the coxless fours, described on BBC Radio 4's *Tweet of the Day* how he saved a one-day-old sparrow chick by feeding it scrambled egg. Sparky became a pet and remained faithful to his master's call even after release into the wild. Lance Corporal Francis Ledwidge, 5/Royal Inniskilling Fusiliers, wrote of the sparrows on the western front in the First World War:

From 'To a Sparrow'

The Venerable Bede's citing of a sparrow, entering from the wintry blast to fly un-noticed above the king and his thanes at their revels in the warmth of the banqueting hall, and then out again into the stormy night, remains an ultimate metaphor of life's chilling brevity.

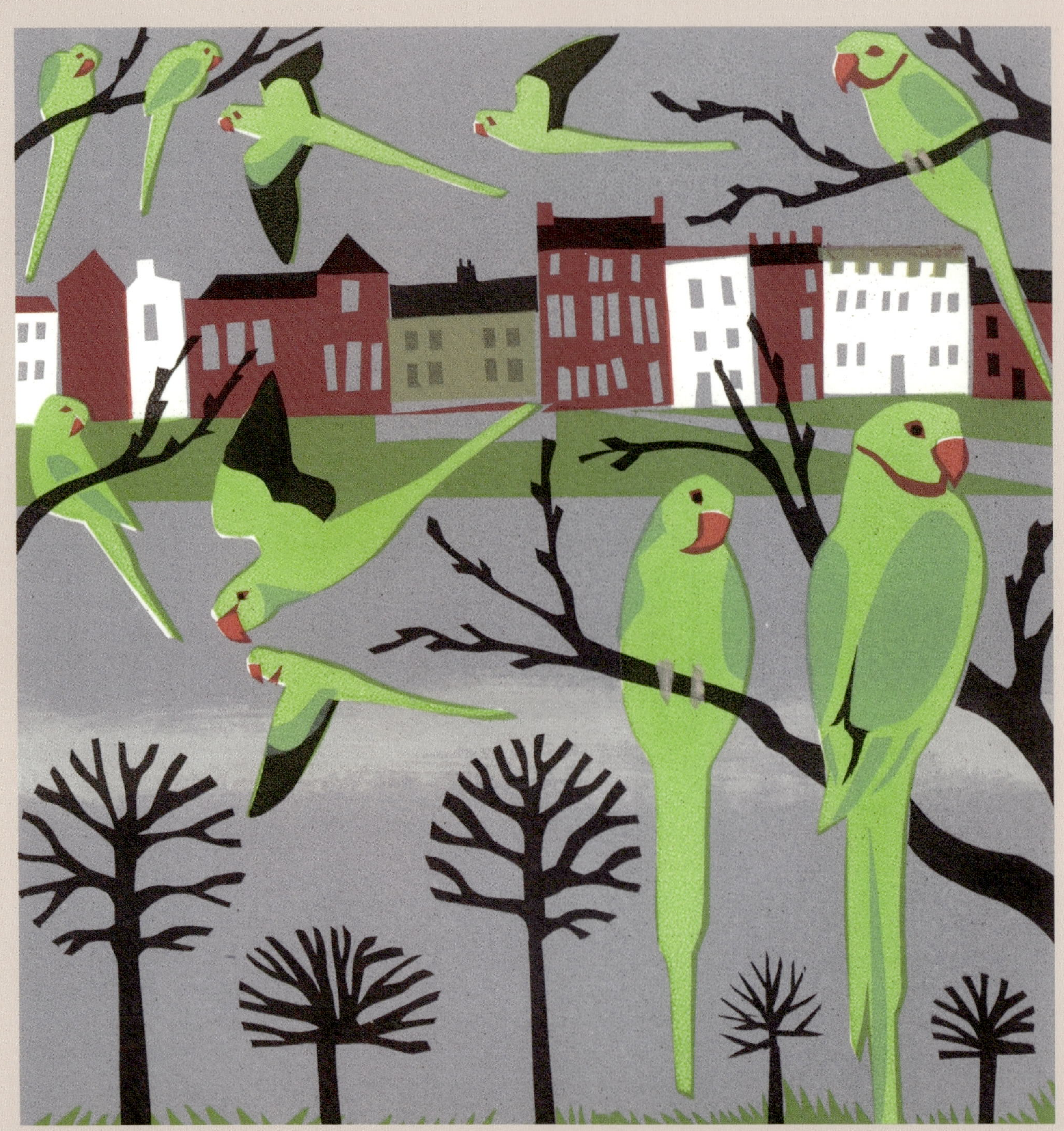

THE ROSE-RINGED PARAKEET

The rose-ringed parakeet (*Psittacula krameri*, named after the eighteenth-century Austrian naturalist Wilhelm Kramer) is also called ring-necked. Only the male has the black, rose-edged ring and black bib. Both sexes have an unshapely rose beak and are vivid green with seven-inch (18-cm) tails. They would brighten any Christmas tree.

It is the most northerly of the world's parrots and probably the most numerous. It was admitted to the British List in 1984, but our acquaintance with this southern Asian and sub-Saharan bird has a much longer history.

Its beauty and mimicry have made it a popular pet since antiquity. It was easy to understand sitting at lunch in John Stefanidis' Belgravia dining room, idyllically placed between courtyard and garden. Framed in the window as a centrepiece were his bird-feeders, momentarily host to two parakeets, a greater spotted woodpecker and two goldfinches: a more jewel-like avian arrangement hard to imagine. It was easy to understand that in India, the parakeets' country of origin, they were prized by kings. Interesting also to learn from Paul Clements, butler, sublime chef and Royal Marine, who minds the feeders, that this little oasis at the heart of London has been visited by 32 bird species, including a mallard.

The rose-ringed parakeet arrived in Europe in the Middle Ages as the most exotic of caged birds. There is ample evidence it reached England: two are carved on a fourteenth-century misericord in Wells Cathedral; another is perched on the Christ child's wrist in a fifteenth-century English book of hours. It was believed parrots automatically said 'Ave' (Hail), the Archangel Gabriel's greeting to Mary at the Annunciation, and were thus symbols of purity. John Skelton (1463–1529), in his satirical poem 'Speke Parott', described his rose-ringed narrator:

Wythe my beke bent, and my lytell wanton iye,
My featheyrs fresshe as ys the emrawde grene,
Abowte my necke a cerculetll lyke the ryche rubye,

My lytell legges, my fete both fete and clene,
I am a mynyon to wayte apon a queen;

The surprise is that such an adaptable species took so long to spread its range. Today it is found from China to the USA; among European cities, it has colonised Rome, Lisbon, Paris and Brussels where, in 1974, it was deliberately released by a zoo owner.

'Postie the Parakeet' – so called because posted through the letter-box of a closed Nottingham RSPCA charity shop one evening in late November 2018 – demonstrated the adaptability of the species. Not only did it survive its ordeal as a gashed bird but it proved the length of a parakeet's breeding season, being only a few weeks old.

In modern England the first escaped birds were seen in Dulwich Park in 1893; the first wild breeding pair in 1930 in Epping Forest; the next not until 1971 in Croydon and Shirley. Then numbers soared; the percentage increase among recent colonisers only surpassed by the little egret. Its squeaky-toy call can be heard throughout England's southern counties and in most of her principal cities, from Bristol to Tyneside. In Scotland, it has reached Edinburgh and Dundee, and one was seen in Cork, south-west Ireland.

Fears for native hole-nesting birds seem unfounded, and greater spotted woodpeckers give it short shrift. Nevertheless, this gaudy bird and its occasional canary-yellow 'albino', may prove unwelcome immigrants. They de-bud fruit trees and are addicted to grapes, thus threatening southern England's burgeoning vineyards. In India, they are a national pest, particularly excoriated for stripping mango groves of their flowers. In Kew Gardens, numbers are already discreetly controlled by shooting.

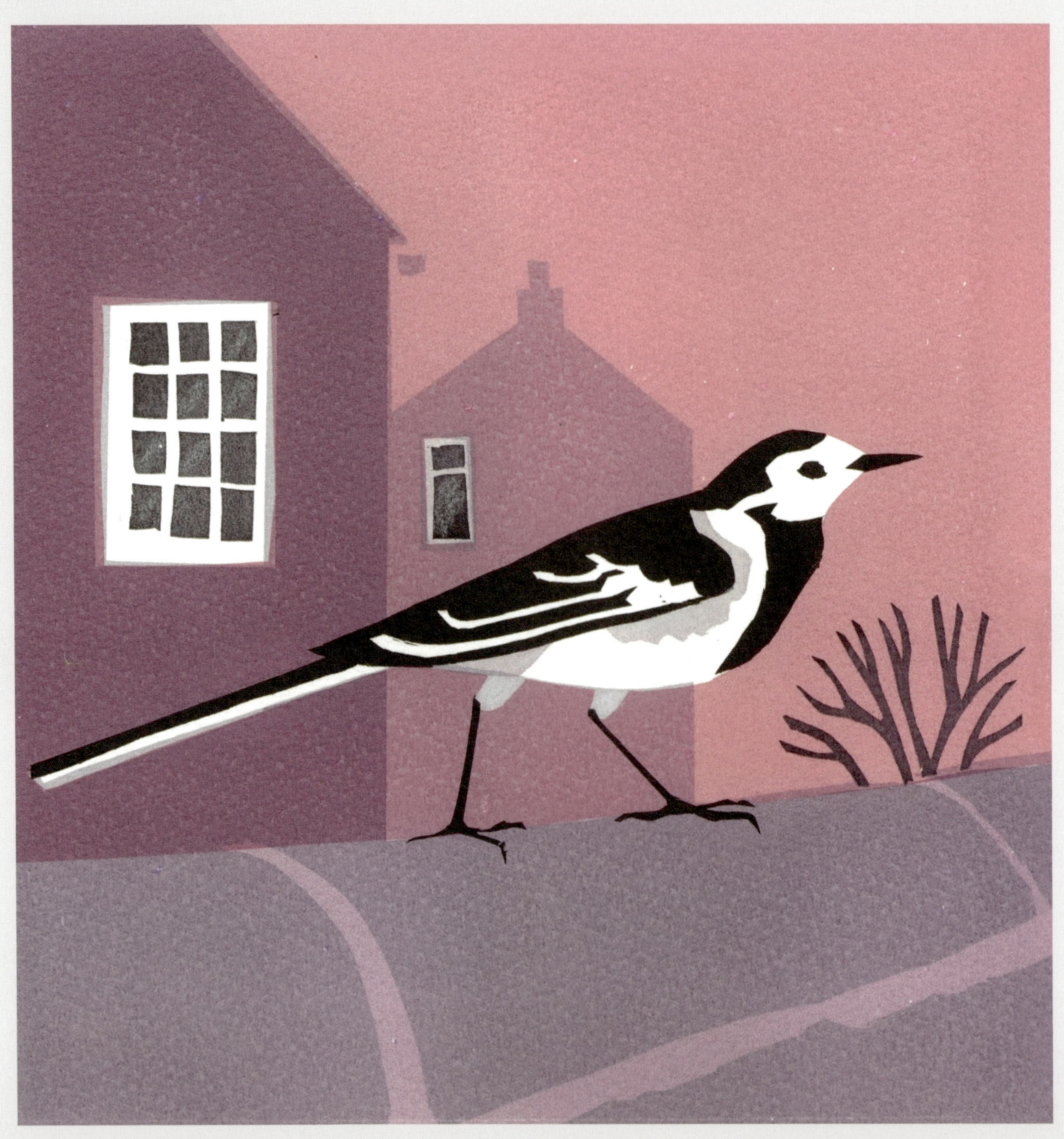

THE PIED WAGTAIL

A mallard pair waddling along Piccadilly was nothing compared with the pied wagtail (*Motacilla yarrellii*) at my feet in London's rush-hour crowd hurrying to the Caledonian Road Underground station. Such a slight bird, so easy to crunch underfoot, darting and threading a passage this way and that in search of the seeming invisible. Pieds constantly surprise. The next one I saw was on an encouragingly plastic-free sandy beach on the west coast of County Cork in Ireland. It was darting and jinking between parents, taking a welcome bask in the sunshine, and their pack of children running hither and thither playing 'bulldog'. The wagtail was equally at ease in either group. In 2018 one even appeared on Wimbledon's Centre Court, during a Djokovic match no less.

Wagtails do wag their tails – nobody knows why, but it gives them a perky air, pavement hoppers not least.

When I walk forth, near me I'd like
Thy pranks to cheer my sorrow.
C. L. BRAIN, from 'The Wagtail'

The nineteenth-century Northamptonshire poet John Clare typically nails the jizz of the bird with his much-anthologised 'Little Trotty Wagtail', intentionally written, it should be said, for children:

Little trotty wagtail he went in the rain
And tittering tottering sideways he ne'er got straight again
He stooped to get a worm and look'd up to catch a fly
And then he flew away e're his feathers they were dry

The pied is still described in *Collins Bird Guide* as 'usually found in cultivated land near water and houses'; and a surprised 2018 observer in Edinburgh wrote: 'As a Highlander, I have always viewed them as water-birds.' Its ability to adapt to urbanisation has armed it against change more successfully than Britain's other wagtails – the grey, a permanent resident like the pied; and the summer-

migrant yellow, which winters in sub-Saharan Africa. The susceptibility to diminished habitat of the grey, a bird of streams and rivers, and the farmland yellow, is reflected in their declines. The present century has seen the yellow's UK numbers halved, the grey's diminished by a third, the urbanised pied's by 10 per cent.

Motacilla classifies wagtails and pipits. The pied is a variation – confined mainly to the British Isles – of the continental white wagtail (*Motacilla alba*), a rare greyer-plumaged passage migrant, which in the twenty-first century, unlike the pied, has colonised the Channel Islands.

A winter-roost survey of the pied wagtails of Poole Harbour in Dorset (2014/15) showed a variety of preferences, from an oil-gathering station to trees around a Tesco car park and yachts in a marina. The heat provided by city centres (light-festooned trees a popular overnight perch), filling stations and airline terminals (notably Heathrow) attracts roosts of up to 4,000.

The cock can string its penetrating 'dinks' and 'chiz-ziks' into a linked song. 'I hadn't heard one sing before, not like this, merrily, even giddily, and not in November,' wrote Richard Smyth (*A Sweet Wild Note*).

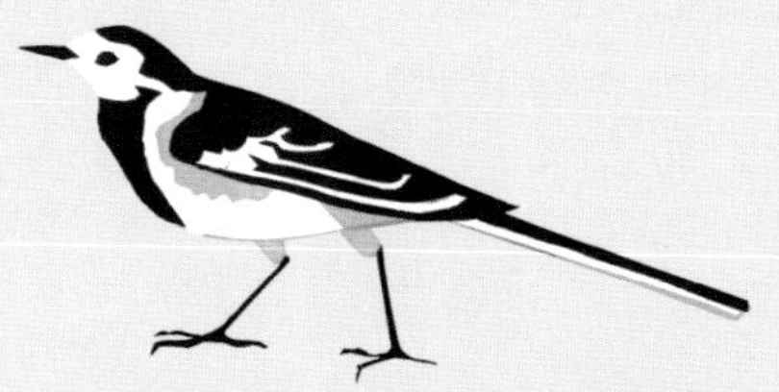

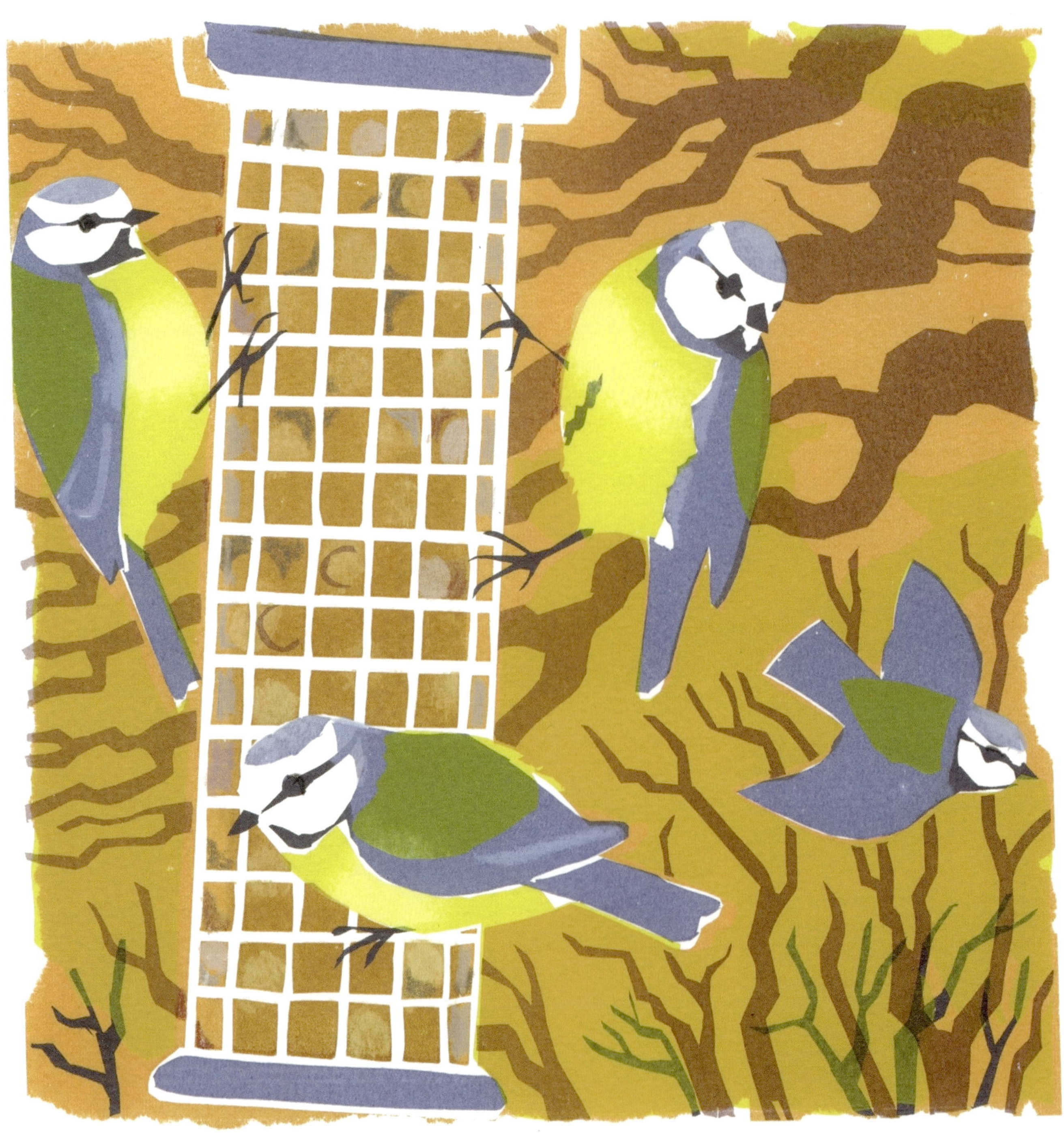

THE BLUE TIT

The playwright William Douglas-Home opened his 1977 *Spectator* review of *The Birdman*, the bird memoirs of his brother Henry Douglas-Home, sportsman, conservationist and pioneer ornithological broadcaster: 'There was once a splendid doctor, with a red beard and a wooden leg, called Clifford, in East Meon, where I live. He came to see my wife, one spring, when she had the flu… "You've got the same thing as Mrs Chubb's got at the bottom of the hill," he told us. Then he added conversationally "What's more, she's got her tits in the letter box again this year."' The Douglas-Homes winced but rallied to acknowledge her good fortune.

Nest boxes, so popular with blue tits, were a nineteenth-century German introduction taken over by the British. Our related obsession to feed songbirds dates from the cruel winter of 1890/91, when British newspapers urged readers to save the starving avian population. The UK spends £200 million pa (2019) on bird food alone, more than the rest of Europe put together. Of this, the blue tit (*Cyanistes caerulus*), most numerous of the titmice tribe at 3 million, is prime beneficiary.

No hobby, popular in living memory, is more reprehensible than egg collecting, of blue tits not least.

It was a ritual exercise in care —
to blow out a finch's or blue tit's egg,
transforming the patterned container of life
into the prize itself. First,
the pin-pricked hole in each end, then
holding it poised to your lips
with nail-bitten fingers and thumbs
like a miniature musical instrument
you were trying to gentle a note from,
pursing your mouth with precise pressure
to start the albumen's gossamer
lengthening into the toilet bowl, And after,
if the egg was fresh as it should be,
the pumping gold of what would now

be no singing bird, in small rich gouts
sinking through the water to the bottom.
And you'd flush that voice away.

GERRY CAMBRIDGE, 'Blowing out an Egg', *aged 12*

Birds need extra calcium to form the shell of their eggs. This is not easy to find in their normal diets, and most seem to depend on snail shells for calcium. The blue tit is especially taxed, laying up to 16 eggs per clutch. For each of its brood, the female blue tit has to find more calcium than forms its skeleton.

It was at Swaythling, Southampton, in 1921 that the Einstein of blue tits pierced or prised the cap of a doorstep milk bottle left by the daily milkman thus adding digestible and energising cream to its diet. Blue tits lack enzymes to digest the lactose in milk; cream is lactose-free. Imitation meant other blue tits, other tit species and birds – starlings, sparrows, blackbirds, magpies – benefited; but by 1949, blue tits still accounted for 60 per cent of incidents. The bonanza ended in the 1980s with the rise of health consciousness and supermarkets. Crucially, milk was often skimmed of its fatty cream.

The RSPB's 2018 Big Garden Birdwatch (see The House Sparrow, page 19) showed the blue tit had risen from fourth to third in the UK list. The previous poll had revealed a 10 per cent decline suffered by all the titmice because of the exceptionally wet breeding season in 2016, which drastically reduced caterpillar numbers, the staple diet of parents and nestlings alike. As a baby blue tit can eat 100 caterpillars a day and there can be 16 of them in a brood, it is easy to see how a shortage of caterpillars is cataclysmic. Nest-box cameras have revealed another predatory enemy in the surprising form of the great spotted woodpecker.

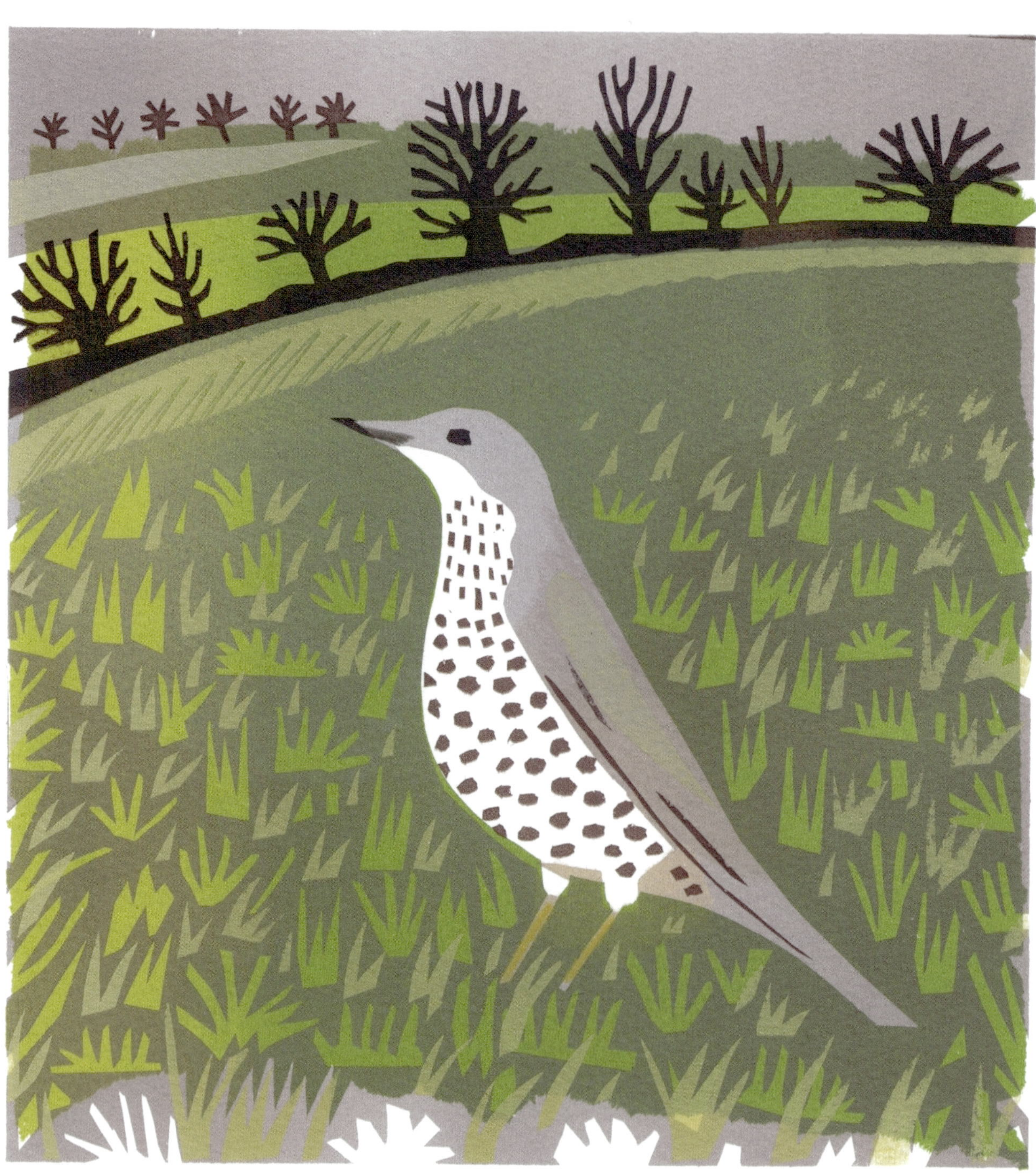

THE MISTLE THRUSH

The most usual call of the mistle thrush is an extended harsh, high-pitched rattle but poets have been inspired by its unseasonal song. In the 'year-cycle charts' James Fisher provided in his British *Bird Recognition* books, he described the song thrush as a winter singer except in rough weather with fullest singing in March–July; whereas the mistle thrush (*Turdus viscivorus*, 'mistletoe gorger'), almost three inches (7cm) longer than its brown, less boldly spotted, more tuneful cousin, sings irregularly from mid-November, with display and song from the winter equinox until June. Moreover, as its colloquial name 'Stormcock' shows, it chooses to sing from a treetop whatever the weather. For that reason, the bird in the popular Christmas poem, Thomas Hardy's 'The Darkling Thrush', is generally identified as a mistle thrush (see Simon Armitage and Tim Dee's *The Poetry of Birds*). *Winter's dregs made desolate* when:

> At once a voice arose among
>> The bleak twigs overhead
> In a full-hearted evensong
>> Of joy illimited;
> An aged thrush, frail, gaunt, and small
>> In blast-beruffled plume
> Had chosen thus to fling his soul

And Edward Thomas wrote:

> I hear the thrush, and I see
> Him alone at the end of the lane
> Near the bare poplar's tip,
> Singing continuously.

> Is it more that you know
> Than that, even as in April,
> So in November,
> Winter is gone that must go?

> From 'The Thrush'

It was the seventeenth-century polymath Sir Thomas Browne who gave the mistle thrush its name. Yet, like the song thrush, apart from miscellaneous berries (which it guards fiercely) it is partial to worms and will similarly bash snails on an anvil-acting stone to break the shells. Toxic to humans, mistletoe berries were an ingredient of trappers' bird lime, thus often the mistle thrush's ironic downfall as much as its sustenance.

Their untidy nests can appear in odd places, such as a red traffic light in Beeston, Leeds, which made the news in 2010. Usually, it is a wary country bird only seeking town in harsh weather, although autumn flocks can briefly alight in city parks. With winter approaching they pair again.

> Raising his voice as the gusts still roared
> A speckled mistlethrush told me clear,
> That harvest was garnered, that apples were stored,
> That summer was ended, that autumn was here.

> BILL HUMPHREYS, from 'Stormcock'

In 'The Darkling Thrush' Hardy also finds consolation in its Christmas 'carolings' defying the blast, even hope of salvation:

> So little cause for carolings
>> Of such ecstatic sound
> Was written on terrestrial things
>> Afar or nigh around,
> That I could think there trembled through
>> His happy good-night air
> Some blessed Hope, whereof he knew
>> And I was unaware.

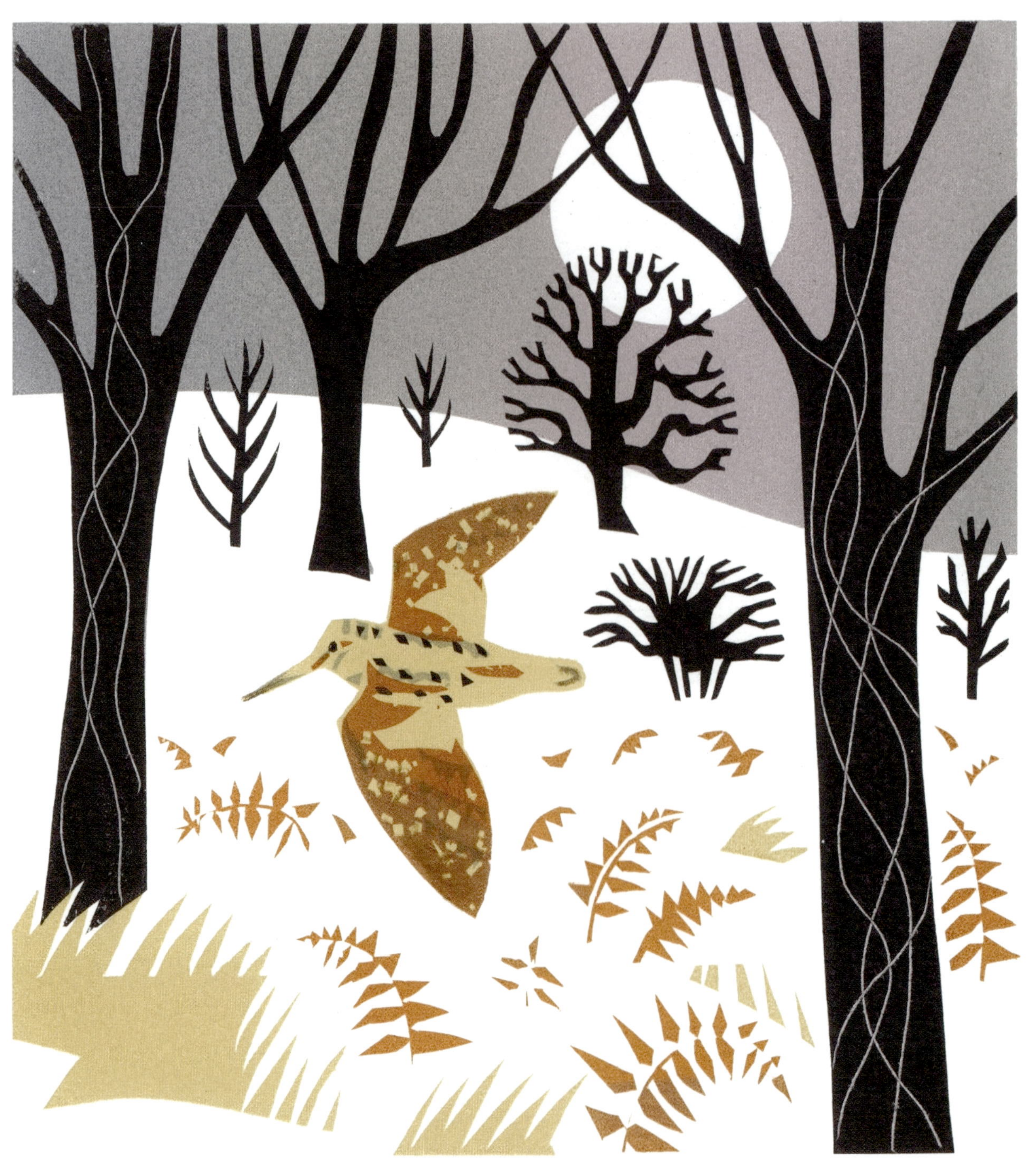

THE WOODCOCK

W. H. Hudson (1841–1922) novelist, naturalist and ornithologist, confirmed that the 'very handsome' woodcock (*Scolopax rusticola*) is 'a favourite alike with the ornithologist, the sportsman, and the lover of delicate fare' (*British Birds*). The woodcock's renown as a delicacy is illustrated in the thirteenth-century *Bird* Psalter and other medieval prayer books. Shakespeare refers to 'four woodcocks in a dish' (*Love's Labour's Lost*, Act 4, Scene 3). The bird is cooked with innards intact adding savoury flavour to the meat. The cooked innards are spread on buttered toast with the roasted bird placed on top. To gourmets' disgust, European law now decrees marketed woodcock must be gutted.

Woodcock flock to migrate but are naturally solitary, nocturnal, birds subject only to the weather. Hence the amusement when the post-Second World War property tycoon, Sir Charles Clore, ordered his head gamekeeper to provide more of them for his guests to shoot.

To adorn a bag of driven pheasants with a woodcock is a feather in the marksman's cap; to miss one, a social gaffe. Lord Clark, the twentieth-century art panjandrum, recalled this sporting ordeal in his autobiography, *Another Part of the Wood*: 'My most dreaded moment was when the cry of "woodcock" went up from the beaters…They are extremely hard to hit…and they had a habit of flitting…down the line to where I stood, shivering and numb.' Lord Home who, as Sir Alec Douglas-Home was the best naturalist and sportsman to be prime minister, shot a woodcock, which fell into the tender of the passing *Flying Scotsman*. He managed to alert the driver, who left him the bird at Berwick-upon-Tweed station with his compliments (*Border Reflections*).

Woodcock have been official UK residents since the 1820s. The 2019 population of 100,000 is enlarged by a winter migration of 1.5 million from northern Europe and Russia. Over 90 per cent shot annually in the UK are migrants. Opposing winds as they cross the North Sea can cause whole flocks to drown from exhaustion. The November full moon, once thought to trigger the migration or 'fall', is called the woodcock or hunter's moon. Until the nineteenth century they were commonly supposed to be lunar migrants:

A bird of passage gone as soon as found
Now to the moon perhaps, now underground.
ALEXANDER POPE, from 'Epistle to Sir Richard Temple'

Satellite telemetry reveals that temperature dictates the exodus. A cold snap usually ensures an influx of woodcock, which then move to the milder west of Britain as they search for penetrable feeding grounds; 50 to 100 are spotted annually in London during the migration months.

Woodcock have to be almost trodden on to be flushed from their nests; only the large and berry-bright eye, unusually large to aid nocturnal viewing, betrays their leaf-litter camouflage. If forced, the female can carry a chick or chicks to safety between its thighs or even on its back. The time to see them is when mating males go 'roding' in the spring twilight. They make a regular, unhurried, circuit of a wood, combined with a repeated ruminative grunting followed by a penetrating 'twiseek'.

As Lord Home wrote, the 'audible "flip" as it jet-propels itself from the ground' is what usually alerts one. It was a wintry woodcock day which stood out in the memory:

'There was a howling northerly gale, and the thermom-eter hung around zero, but the woodcock were "in". The cold snap had been sudden, and the ice-crust on the bogs was thin so that our feet by lunch-time were in danger of frostbite. Somebody suggested pouring whisky into our boots…For the rest of the day, we walked on air!'

They are regularly seen in London. Alister Warman, when he was director of the Serpentine Gallery, flushed one in Kensington Gardens. A decade later, when principal of the Byam Shaw School of Art in Holloway, north London, a stunned bird was brought to him by a student. It was later released in the jungle of Old Highgate Cemetery, ideal woodcock terrain.

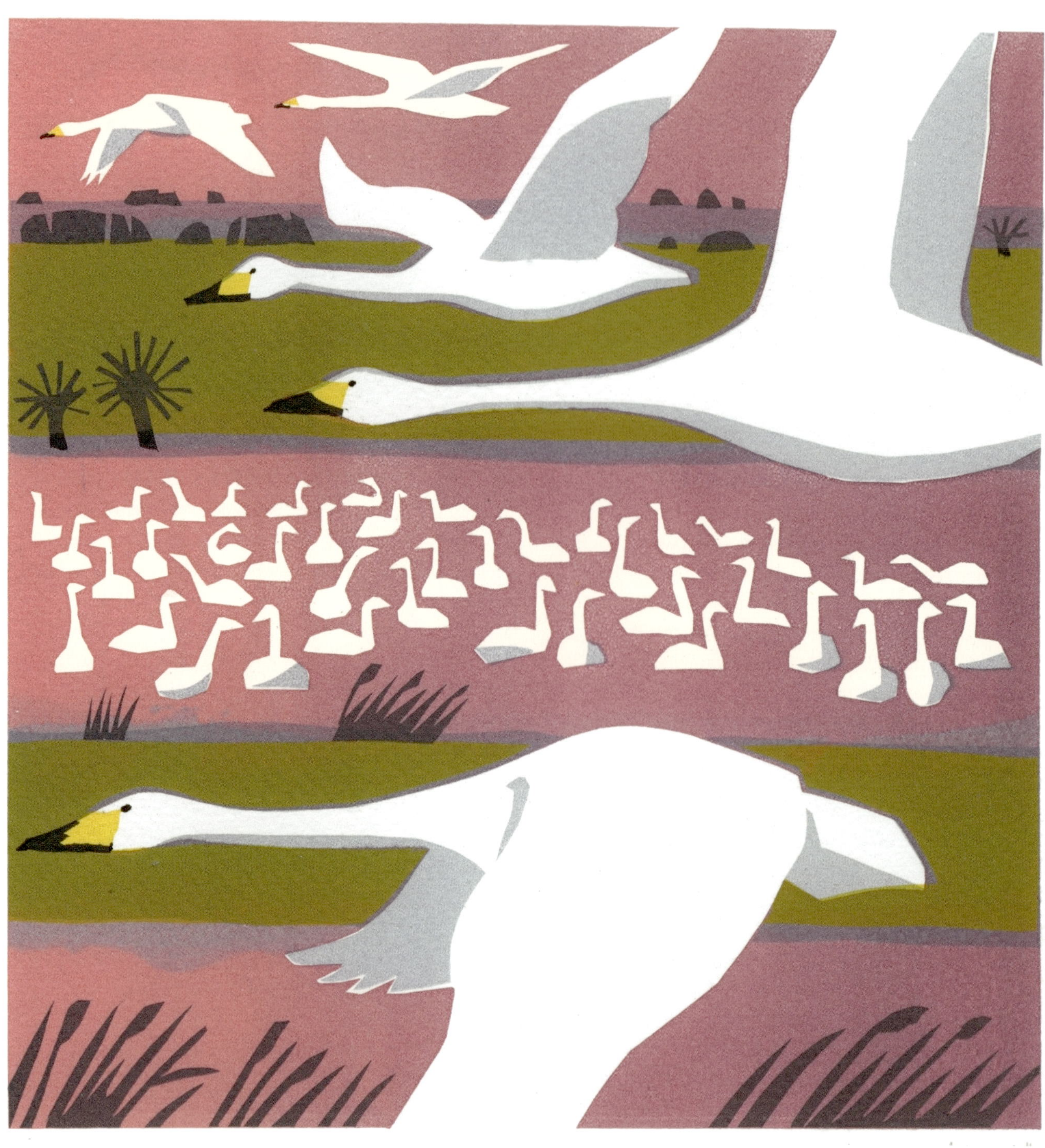

THE WHOOPER SWAN

I have looked upon those brilliant creatures,
And now my heart is sore.
All's changed since I, hearing at twilight,
The first time on this shore,
The bell–beat of their wings above my head,
Trod with a lighter tread.

W. B. YEATS, from 'The Wild Swans at Coole'

The swan seen in Britain throughout the year is the semi-tame, orange-billed, voiceless, hence mute swan (*Cygnus olor*). In winter two clamorous wild swans – the black-and-yellow-billed whooper (*Cygnus cygnus*) and Bewick's (*Cygnus bewickii*) – migrate here. Until the nineteenth century all wild swans were 'hoopers', the sonorous bugle call of the whooper, from the Anglo-Saxon 'whopan' to 'cry out'. 'Swan' is itself an Old English word for 'sound'. Swans, mating for life, symbolise fidelity. It adds poignancy to the myth of the 'swansong', when the bird is said to produce a flute-like wailing with its dying breath, a sound that may be unpoetically explained by the whooper's extra-long trachea.

The whooper has a global range from Ireland to Japan, but ours migrate from Iceland. When Yeats published his poem in 1917 the wild swans at Coole in County Galway could have been Bewick's, but today they would be whoopers, since the Bewick's shrinking population no longer extends to Ireland. The longer-necked whooper is the more majestic: a fifth larger, with a wingspan up to nine feet (2.7m) and a record weight of 34 lbs (15.4kg), which makes it one of the heaviest flying birds. It can withstand -50°C, and a migrating skein over Northern Ireland was seen by an airline pilot at 29,000 feet (8,848km), the height of Everest. This was confirmed by radar, the fourth-greatest recorded height by a bird; the more impressive for being unaided by Andes or Himalayas. Radio-tracking reveals they prefer to hug the sea on migration, enabling settling rests. They can complete the 600 miles (966km) from Iceland in under a day. The wing beat of the mute throbs; the whooper's is a rhythmic whoosh.

The recently booming Icelandic migration (34,000) means Britain hosts a fifth of the global population, the largest concentration (9,000) at the various winter-flooded reserves of the Ouse washes (Cambridgeshire/Norfolk). Their 'constant racket' always makes Carry Akroyd wonder 'what the resident mutes think of all the noise arriving'. The increase is mainly attributed to milder Icelandic springs and ample winter fodder; like pink-footed geese, they have developed a taste for sugar-beet tops. They rarely breed in Britain.

Wild swans are enshrined in European folklore and fairy tales as good omens. In the Highlands and Islands they were regarded as angels, to the fury of a pre-Second World War correspondent to *Shooting Times* who wrote that in the interests of sport such nonsense should be dispelled 'once and for all' (*Birds Britannica*). They have been protected in Britain since 1954, across Europe since 1979.

But now they drift on the still water,
Mysterious, beautiful;
Among what rushes will they build,
By what lake's edge or pool
Delight men's eyes when I awake some day
To find they have flown away?

W. B. YEATS, from 'The Wild Swans at Coole'

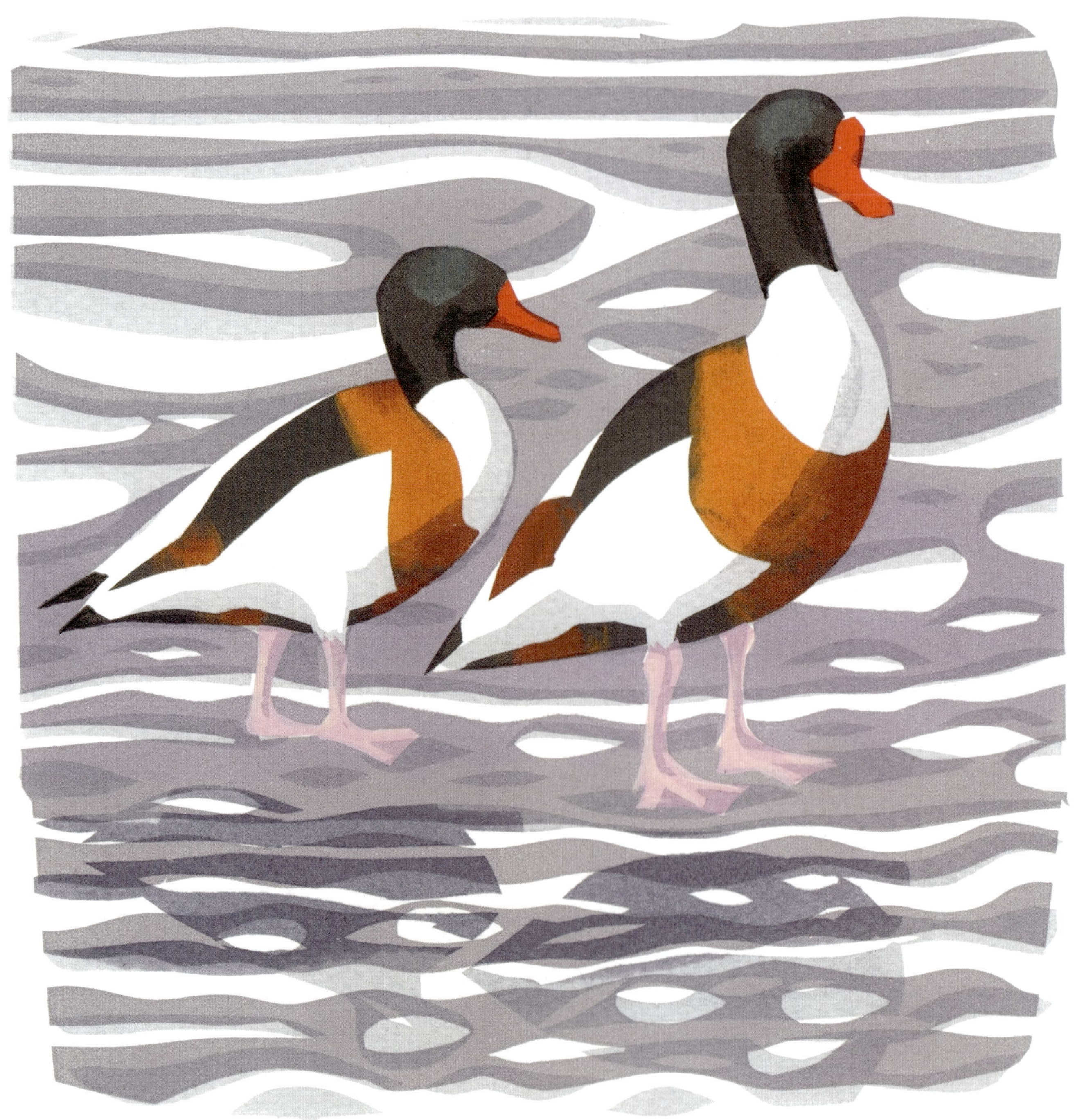

THE SHELDUCK

How often a coastal train journey is made memorable by the flag-like glimpse of a shelduck (*Tadorna tadorna*) along the tide line or far out on mudflats. It is the pied plumage, exceptionally among ducks the same for male and female, which gives its name (from Dutch *sheld*, 'variegated'), although molluscs form a large part of its diet. Shelduck are sturdy (2 lbs/900g), sedate, slow-winged birds and, although classified as a duck, are placed in bird guides among the geese. They have none of the low-slung, streamlined and flying urgency of duck. Colloquial names such as skeel goose, links goose, ringer goose long ago made the association.

The predominance of white is what catches the eye. A closer view reveals the striking arrangement of rust breast band, back and under-tail; bottle-green head and wing flashes (specula), black pinions, sealing-wax scarlet bill and longish pink legs, which they stamp to expose lugworms. Perhaps it is the prominence of the muscovy-duck-like knob at the base of the drake's bill which makes it seem a decorative farmyard bird out of its element. It lacks the poetry of a wild bird, although wild it is; and for all its flamboyance it has been disdained by poets and wildfowlers. Its unmusical call is a laughing 'ag-ag-ag' and its meat, for the sporting naturalist Brian Vesey-Fitzgerald (1900–81), tasted of paint. When a Norfolk warden was asked by a journalist the best way to cook a shelduck he thought the man was taking the mickey, so he facetiously answered in kind: place the duck in the oven alongside a brick, he told the journalist, when you can get a fork into the brick the bird will be done. His joke 'recipe' was duly reported as a genuine piece of country lore (*Birds Britannica*).

W. H. Hudson, a founding member of the Royal Society for the Protection of Birds (RSPB), is exceptional in having a soft spot for shelduck. He described them having a 'strange guinea-pig arrangement of three colours – black, white and red', and was especially amused by the part played by the drake when the duck was laying. 'I do not know any other species in which the male takes it upon himself to instruct his mate on a domestic matter which one would imagine to be exclusively within her own province.' Moreover, he thought he was the first to describe this curious 'sign language'.

The birds would be flocked when one drake would begin to move his head 'like a pianist swaying his body in time to his own music' to cajole his mate into going to her burrow to lay an egg. She would reluctantly follow him. If she stopped, he would start swaying until she obeyed. At the burrow's mouth, she would sometimes make a last stand, as if 'saying, "I have come this far just to please you, but you'll never persuade me to go down into that horrid dark hole"...he prevails; and bowing her pretty head, she creeps quietly down and disappears while he remains on guard at the door – for a little while.' (*Birds and Green Places*)

Ducks that fail to find a burrow sometimes lay their eggs in an existing nest. The unsuspecting hosts can find themselves parenting a double-brood of more than 30 ducklings. Shelduck are notably communal birds. When the ducklings have left the nest, up to a hundred at a time can congregate in 'creches' under the supervision of a single adult pair. Every July a majority of the 180,000 British population responds to a similar instinct for safety in numbers by migrating to moult on the vast tidal flats of the Heligoland Bight, off the German North-Sea coast opposite Lincolnshire. The strategy misfired in August 1954, when the RAF used the area for bombing practice, inadvertently killing thousands of the birds. The Air Ministry recognised its mistake and agreed that in future it would not use live ammunition during the birds' moult period.

Shelduck begin to return in September, swelling the growing ranks of those preferring to stay in Britain for the moult on the Wash, notably at Bridgewater Bay and on the Severn estuary. In winter they are widely distributed.

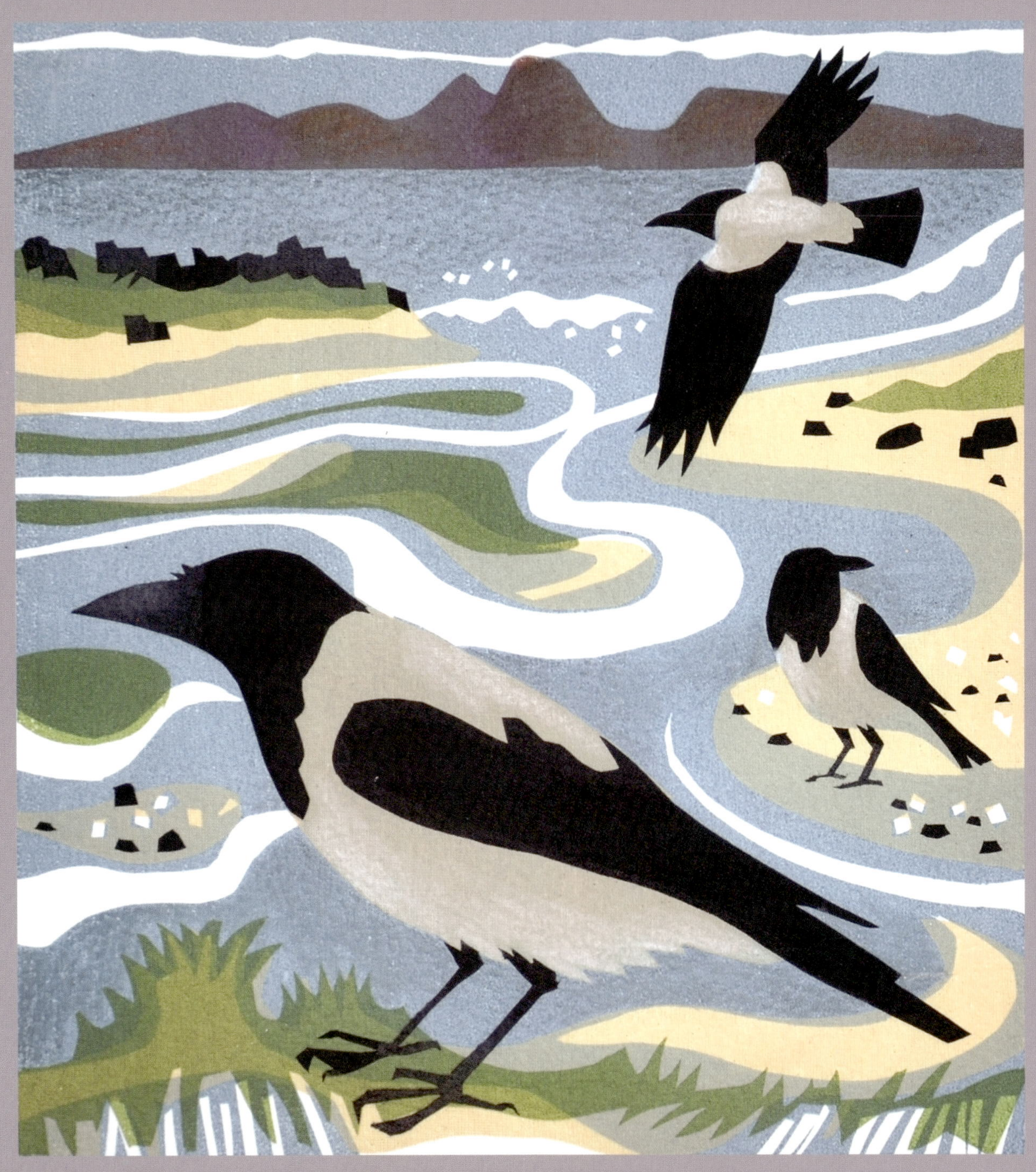

THE HOODED CROW

In 2002 the hooded crow (*Corvus cornix*) and carrion crow (*Corvus corone*) were taxonomically reconfirmed as separate species. The geneticist Peter de Knijff commented: 'I find any discussion about species concepts irrelevant…for birds themselves (or any other organisms) this does not matter.' Quite. The raven is a 'corbie' in Scots but so are both the crows, the carrion also called a 'hoodie', giving 'hoodie', when applied to loitering teenagers, a decided edge.

Sudden outside the high window, one crow
    Hangs in the air
And lights on a withered oak–tree's top of woe.

One bird, one blot, folded and still at the top
    Of the withered tree! — in the grail
Of crystal heaven falls one full black drop.
    D. H. LAWRENCE, from 'In Church'

The carrion is black, the hooded grey but for head, crop, wings and tail. In Britain, the 'socially dominant' carrion rules in England, Wales, the Scottish lowlands and north-east; the scarcer hooded in the Isle of Man, the north-west and islands of Scotland – Carry Akroyd shows them against distant Jura – and Ireland, where a carrion sighting is newsworthy in the Republic. Where they intermingle, they sometimes cross-breed. On mainland Europe they divide west and east. Sun-seekers will find carrions in Spain and France, the hooded in Italy and Greece. Nesting sites and feeding habits are determined by habitat. Unlike rooks, crows are solitary nesters. I am indebted to Richard Roberts, film-maker, gardener, novelist, for the insight that Shakespeare's knowledge of birds adds further metaphorical weight to the key lines in *Macbeth*, Act 3 Scene 2:

Light thickens, and the crow,
Makes wing to th' rooky wood.

Macbeth, in crow-like isolation through his crow-like misdeeds, must nonetheless follow his ambition to be king in the gregarious rookery of the world, and thus complete his tragedy.

Both crows are solitary nesters, but the hooded's largely treeless British habitat forces it to use the ground or cliffs. Similarly, it is more likely to feed on shellfish, which it cracks ingeniously by dropping them from a height onto rocks.

Crows are the bird world's undertakers, black portents of doom as in *Wheatfield under Threatening Skies with Crows*, painted by Van Gogh the month he committed suicide.

Crows and parrots are the most intelligent bird groups; crows one of the few species with the brains to use implements. In October 2018 Mango the carrion crow had to plug three tubes together to recover an otherwise unreachable morsel of food provided in a box. The feat, recorded by researchers from Oxford University and Munich's Max Planck Institute of Ornithology, was the first time a non-human had put together a compound tool of more than two parts.

Their taste for other birds' eggs makes them the arch-enemy for gamekeepers, whose declining numbers have surely contributed to the near 100 per cent increase in carrion numbers since 1970 – with 200,000 breeding pairs in Scotland alone. Crows are described as solitary, but visit London's Kenwood in winter to find yourself in Hitchcock's *The Birds*. Until the late 1940s there were gamekeepers in London's royal parks. Between 1995 and 2010 the carrion crow population increased by 65 per cent, whereas the increase for the UK as a whole was 10 per cent; meanwhile, songbirds dwindle.

The soft options of roadkill and landfill are a contemporary bonus as far as their predatory victims are concerned. Will it be the last bird left standing? Fly fisherman and poet laureate Ted Hughes thought so.

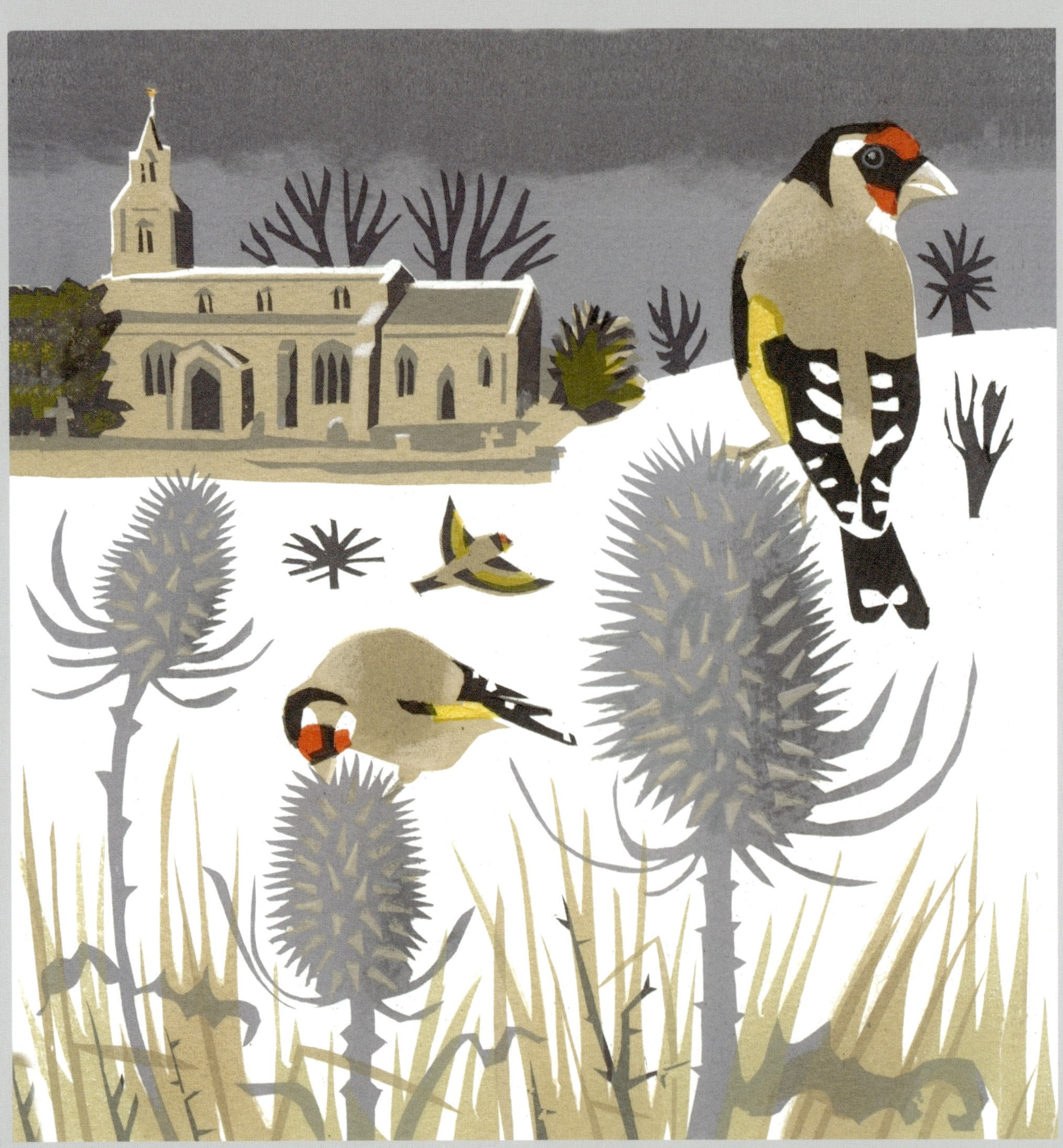

THE GOLDFINCH

Brightly plumaged birds are associated with the tropics, but for prettiness, even a parakeet or hummingbird cannot beat the goldfinch (*Carduelis carduelis*). No wonder a goldfinch flock is called a 'charm', although it is their twittering call, from Old English 'c'irm', rather than their pretty plumage, which inspired the name. In Vivaldi's *Il Gardellino* a flute imitates their charming sound.

Prettiness has cost the bird dear. Most heinously there was Henry VIII's 1532 Preservation of the Grain Act, augmented in 1566 by Elizabeth I and not repealed until the eighteenth century. This put a price on the head of anything in England, feathered or furred, that ate marketable food. It included the blameless goldfinch, which principally feeds on thistle seed. Would this harmless forager have been condemned but for its eye-catching appearance?

Certainly, prettiness was the cause of its next downfall. 'Unhappily, it is now not very easy to see them,' wrote W. H. Hudson in 1895. 'For the goldfinch is a favourite cage-bird, and so long as bird-catching is permitted, this charming species will continue to decrease.' Help was at hand. Thomas Hardy pulled heartstrings with his sad little poem, 'The Caged Goldfinch':

> *Within a churchyard, on a recent grave*
> *I saw a little cage*
> *That jailed a goldfinch. All was silence save*
> *Its hops from stage to stage.*

In 1904 the Society for the Protection of Birds, originally Emily Williamson's anti-featherware Plumage League, gained royal patronage by becoming the RSPB.

The bird was also a favourite pet, those flashing gold wings long making it the bringer of good health and luck. There was a saying that if a goldfinch flew over a bachelor girl, she would marry a millionaire. The upsurge in goldfinch numbers since the mid-1980s coincided with rocketing house prices, which must have made married millionairesses

of many an iconic 'girl in pearls' featured as the photographic frontispiece of *Country Life*.

In his poem, 'The Redcap' (a country name), John Clare wrote that goldfinches 'are never far from town'. This has been especially true since addiction to imported Nigerian niger seed meant UK garden sightings rose from 10 per cent to 60 per cent between 1995 and 2011. Latterly the return of sparrows to my London garden has banished the goldfinches, which Simon Courtauld, *The Oldie* magazine's Wiltshire-domiciled Kitchen Garden columnist, reports, now prefer sunflower seed, confirmed by Michael Chaplin at Cooling in Kent.

Its prettiness has made it a favourite with artists. An outsize goldfinch features in Hieronymous Bosch's masterpiece, *The Garden of Earthly Delights*. Its decorative and symbolic presence in medieval and Renaissance pictures is explained by the legend that those flecks of red were acquired when trying to pick the thorns from Christ's gory crown on the road to Calvary. Its famous portrait by Carel Fabritius, Rembrandt's finest pupil, provided the cover for Donna Tartt's bestselling 2013 novel, *The Goldfinch*, in which the painting plays a central role. The popularity of the book meant the picture was moved from a landing to a gallery in The Hague's refurbished Mauritshuis museum. As a result of the similarly induced popularity of Vermeer's *Girl with a Pearl Earring*, this previously hidden gem of a museum had to be enlarged to cope with the crowds. The increased attendance more than matched the British goldfinch population of 300,000 breeding pairs.

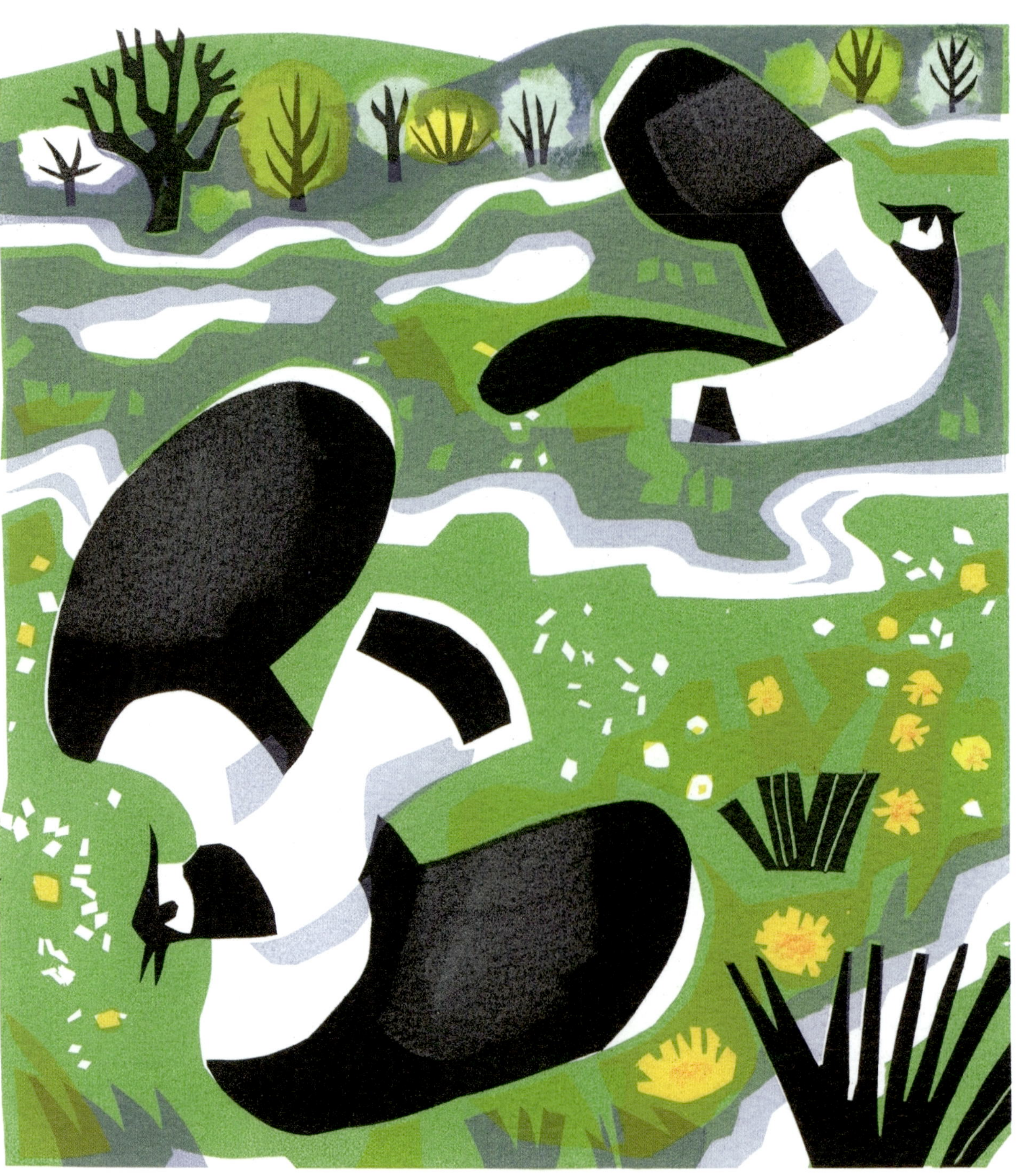

THE LAPWING

The spring flight of the lapwing (*Vanellus* – little fan – *vanellus*) has a flamboyance that matches the season's zest. The males sculpt the air on humming wings with an ecstatic, tumbling, display over the nesting grounds, whether new-sown wheat or upland pasture and moorland; uttering that thin, wheezy call by day and night, perhaps best mimicked by its east-Scottish name, 'peasiwheep'. 'Lapwing' describes its normal flight, as it flops along on those fan-shaped wings, their black and white contrasts lending a flicker to a travelling flock. Yet one derivation is from an Anglo-Saxon word, which does not describe the flight but rather the long, black, curving crest, the most singular feature of its beautiful plumage. Distant black and white, even its secondary name of green plover, give little indication of the apricot tail-coverts, the olive mantle and wing-coverts, shot through with emerald, iridescent bronze, purple and pink.

The collective noun for lapwings is a 'deceit'. 'The false lapwynge, ful of treacherye' wrote Chaucer in *The Parlement of Fowles*. And Shakespeare's womanising Lucio boasts in *Measure for Measure* Act 1: Scene 4:

> … 'tis my familiar sin
> With maids to seem the Lapwing, and to jest
> Tongue far from heart

The reputation unfairly comes from the bird's defence of its young, when it lures away a predator by hirpling off dragging a wing to appear wounded and therefore easy prey. The chicks precociously run at birth, Ben Johnson (*Staple of News*) described rash people as 'lapwings with a shell upon their heads'.

Lapwings were prized for their meat and eggs. Selling plover (lapwing) eggs was a legitimate business in the UK until the 1926 Lapwing Act. In 2015 a Frisian mayor refused a ceremonial plover's egg to the fury of traditionalists, the custom pre-dating the medieval *Elfstedentocht* (skating race). A reminder that the Easter egg hunt and Easter bunny both owe something of their origin to the lapwing; the shared fields of lapwings and hares giving rise to the notion that bunnies were born from eggs.

The Frisian mayor's ornithologically aware refusal reflected the grave fall in numbers – in Britain a two-thirds decline since 1970. Lapwings no longer nest in most of south-west England, western Wales and the west of mainland Scotland; while in Ireland the number has halved. Nesting pairs remain highest in north-west England, the Outer Hebrides, Orkney and Shetland. Modern farming, notably autumn sowing, appears the principal cause of decline.

Winter sees lapwings congregate in flocks, the numbers bolstered, subject to the weather, by influxes from the continent. Sometimes winter flocks are peppered with golden plover, which look contrastingly brown and thrush-like. In recent years lapwings have shown a preference for wintering on or near Britain's eastern estuaries, and rooftop flocks can now be seen in towns.

In faraway Samoa, Robert Louis Stevenson remembered them from his Scottish youth in his sonnet to the Galloway novelist S. R. Crocket:

> *Be it granted me to behold you again in dying,*
> *Hills of home! And to hear again the call;*
> *Hear about the graves of the martyrs the peewees crying,*
> *And hear no more at all.*
>
> From 'To S. R. Crockett'

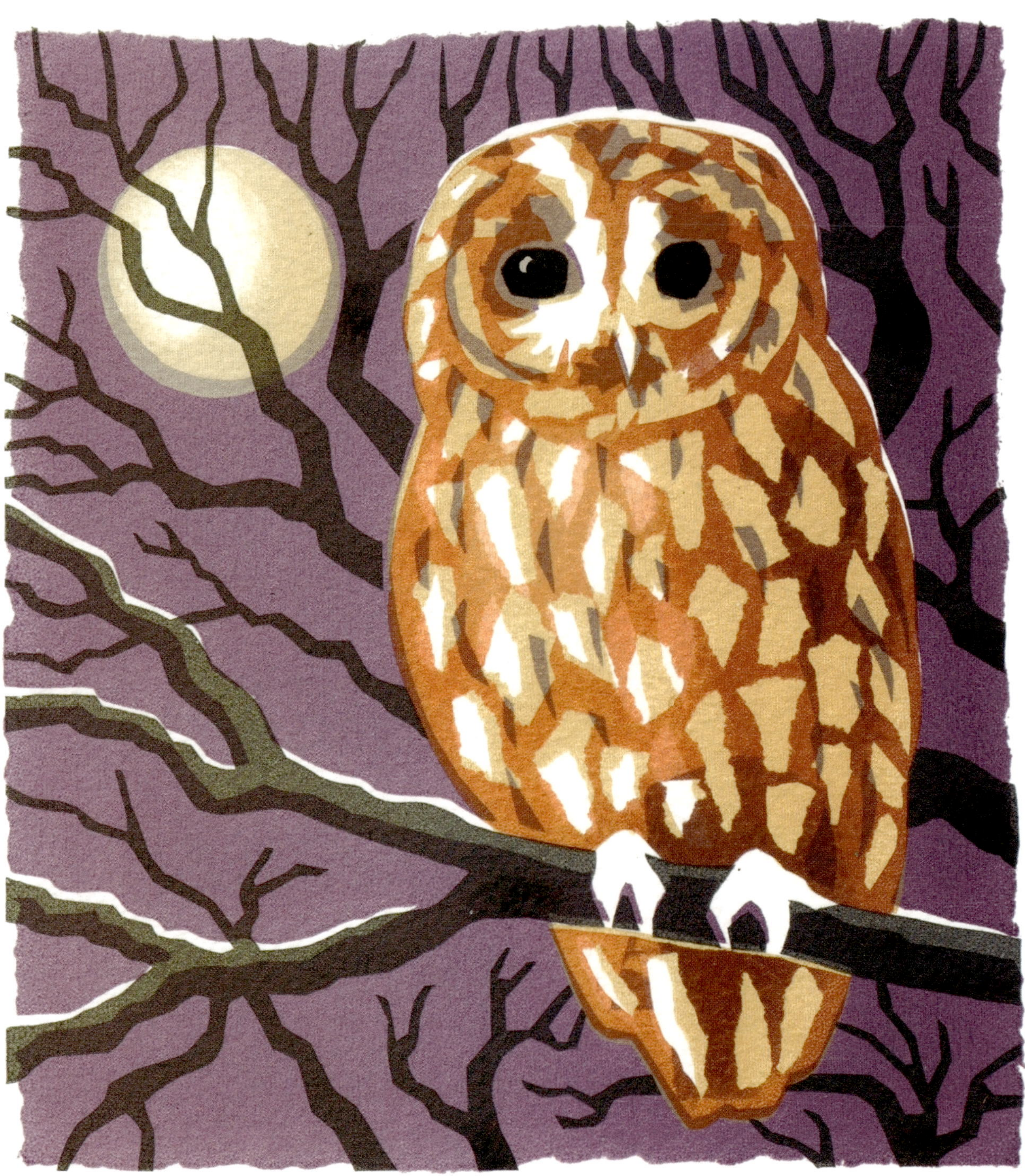

THE TAWNY OWL

One February day I had a shock. There was a porridge-coloured lump in the bare chestnut tree. It was too solid for litter and apparently had head and shoulders. Could it be a sloth escaped from nearby London Zoo? Binoculars revealed a huddle of three young tawny owls. The tawny (*Strix aluco*), with its woodland cousin, the long-eared, is earliest of owl nesters. In 2014 a clutch was reported in the first week of January. Urban tawnies (and other urban birds), encouraged by the greater warmth that cities and towns offer, tend to nest earlier than those in the countryside.

It was not until dusk that a parent bird fed the owlets. Used to tawnies in daylight, sleepily flapping from tree to tree to escape a noisy avian mob led by pinking blackbirds, it was another shock to see the mature bird arrive at the searing speed of a hawk. Easy to appreciate the famous wildlife photographer Eric Hosking's account of losing half his sight to one: 'There was not a sound…The owl, with its night vision, had dive-bombed with deadly accuracy, sinking a claw deep into the centre of my eye.' (*An Eye for a Bird*)

They can make endearing pets. Henry Douglas-Home's most memorable pet tawny was entrusted to him by a cousin. Tawnies can become moribund in extreme cold. She found one 'dead' and was amazed when, after 30 minutes indoors, it woke and laid an egg. Douglas-Home looked after it for her secretly when staying at Edinburgh's New Club, at that Second World War moment, 'bristling' with top brass. At night from the bedrail, it threatened to betray him by answering a rival in Princes Street Gardens. Eventually, he forgetfully left the door ajar, and it felled a startled colonel in the darkened passage. The game was up. 'I was very sad. It had been a trusting character of great charm, entirely incapable of looking after itself.'

Tawnies may look like cuddly toys or the bespectacled oldies of children's books, but Beatrix Potter accurately showed Squirrel Nutkin risked death when he teased Old Brown. Nor is their diet confined to rodents: BBC TV's *Autumnwatch* revealed one taking an adult magpie and half-grown rabbit. At dead of night in Berwickshire, I was woken by one snatching a roosting songbird from the creeper under the bedroom window.

From New Year the males vocally establish their territories, as Shakespeare described in *Love's Labour's Lost*:

> *When all aloud the wind doth blow,*
> *   And coughing drowns the parson's saw,*
> *And birds sit brooding in the snow,*
> *   And Marion's nose looks red and raw,*
> *When roasted crabs hiss in the bowl,*
> *Then nightly sings the staring owl,*
> *   Tu-whoo;*
> *Tu-whit, to-whoo, a merry note,*
> *While greasy Joan doth keel the pot.*

The note elsewhere in Shakespeare is far from 'merry' but an omen of death. For W. H. Hudson it added 'mystery' to a wood. In fact, males hoot but both 'keewick'. They also call by day. Walter Scott compared critics to owls, who like 'to make a noonday darkness'. Henry Douglas-Home once had one of his BBC outside broadcasts sabotaged by his brother William, who, as a practical joke, hooted into one of the unit's woodland microphones. The result was considered a masterpiece of bird recording, and thereafter whenever Juliet appeared to Romeo at her window, she did so to the accompaniment of William's hooting.

Tawnies rarely stray beyond a mile, so one felt for the buffeted bird found in March 2014 at Crick, Northamptonshire, which had clung to the front of a freight train from Glasgow. Also for the bird in the Forest of Dean, Gloucestershire, which was found gazing out through the window of a mercifully unlit wood burner after falling down the flue. Their soft feathers are easily waterlogged, making them prone to drown in cattle troughs, etc., presumably mistaking their reflection for a rival. At 40,000 pairs, tawnies are in decline but remain the UK's most numerous owl. They are unknown in Ireland.

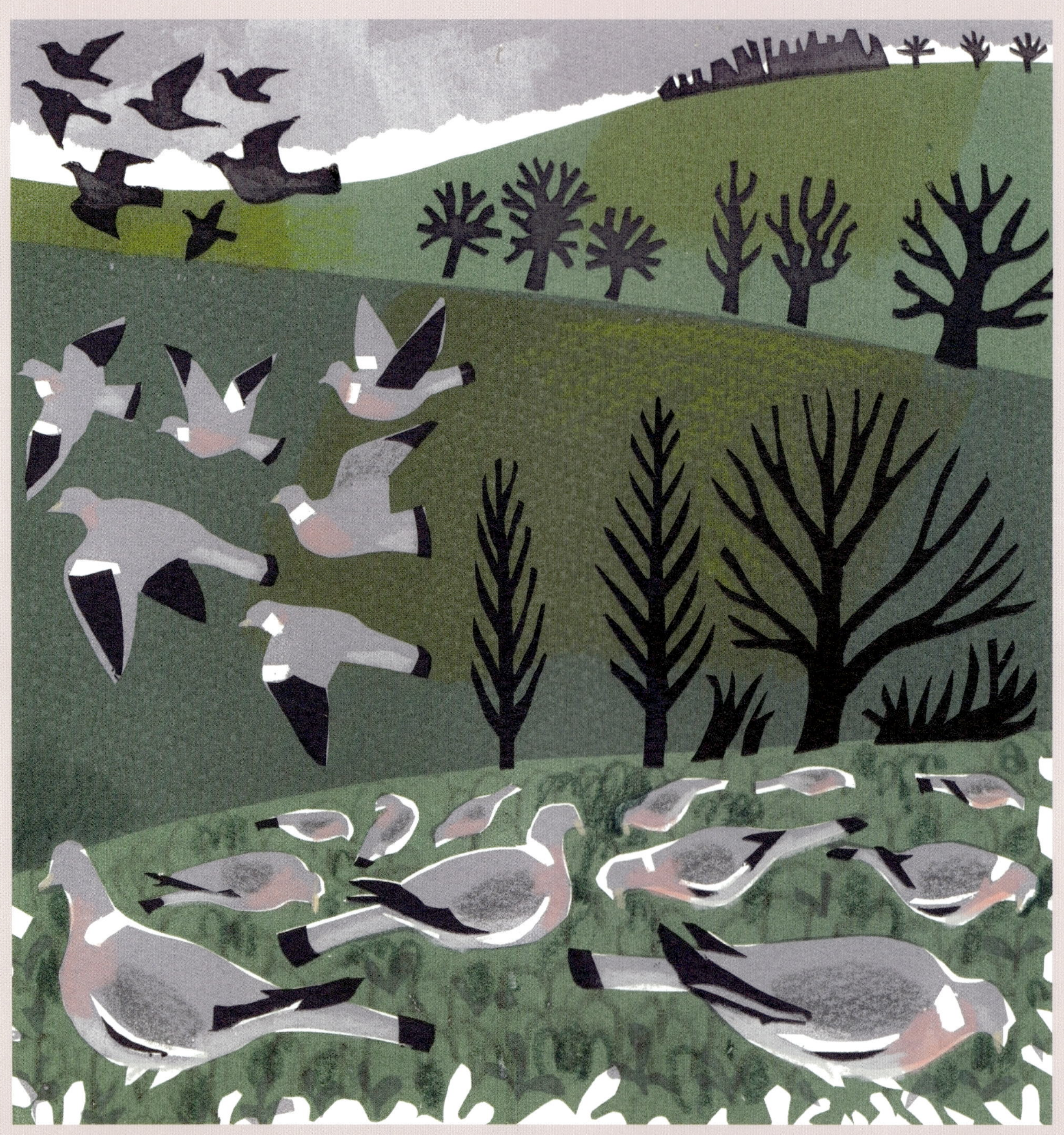

THE WOODPIGEON

NORMAN MacCAIG, from 'Windy Day in March'

The woodpigeon (*Columba palumbus*) is a success story. The UK breeding population increased 169 per cent between 1967 and 2010. There may be 10 million by summer's end, and they breed throughout the year, food permitting. More arable land has helped winter survival. Huge unexplained flocks were seen in Scotland and England in 2016, one of 90,750 counted (even an expert could not say how!) over Painswick Beacon, Gloucestershire. Tame in town, wild in the country, they can be a pest to bird lovers and farmers alike.

Bird tables reveal a bullying side. It takes a starling, with its dagger beak and aggressive gait, to show who really rules the pecking order. Having cleared the bird table, the woodpigeon does its worst to the bird bath. One dozes hunched for hours on the edge of mine, just the tip of its tail in the water. In common with all pigeons (the 300-strong *Columbidae* family) the woodpigeon sucks to drink rather than dip-tilts like other birds.

Its slovenly behaviour can blind us to its charms. W. H. Hudson commends it as the handsomest of British doves. Traditionally it was more poetically called a ring dove, 'its dove-grey tints being singularly delicate, soft, and harmonious'; best appreciated when it flops about, seeking a purchase with spread tail and wings, to feed on a wall of ivy berries. Then one can fully appreciate the neck's white flashes (adults only) and petrol-stain iridescence, the cold black white-edged pinions and white wing coverts – fatal semaphore for its avian enemy, the peregrine.

The breast of a woodpigeon is a cold pink, but sunlight can suffuse it with warmth.

ALAN SWARDEN, 'Waiting for Pigeon'

There is also its soothing call, the coo-cooing ending on an abrupt half-note; in summer dawns many calling simultaneously seem consciously to harmonise into a rhythmic, far-flung chorale. There are exceptions, as a 2017 letter from Bill Sylvester of Rickmansworth, Herts testifies: 'In my garden the pigeons don't "coooo" but continually bombard me with a "Now hear this!!" Uncanny, as for many years I was the "Town Cryer" at Carnival time.'

Females are fair game at any season. As soon as one alights, down comes a male to strut his stuff: rapid pursuit is followed by a courtly bow. It usually ends in flight to a branch, where the ritual is repeated. Eventually, she wings it or submits, dutifully crouching while he hops aboard and has his flapping way. There is also the slow, ecstatic, wing-clapping display flight. The nest, a net of twigs, is seemingly insufficient to bear the two white eggs, let alone a weighty (more than 1lb/454g) bird. There can be six broods – the two-egg limit explained by the squabs need for regurgitated food, a time-consuming feeding method.

In the wild, it is the wariest of birds. Lord Home did not know another that turned quicker 'at the sight of a gun'. The meat makes a rich meal, but have the night-time indigestion tablets handy if it has been served for supper.

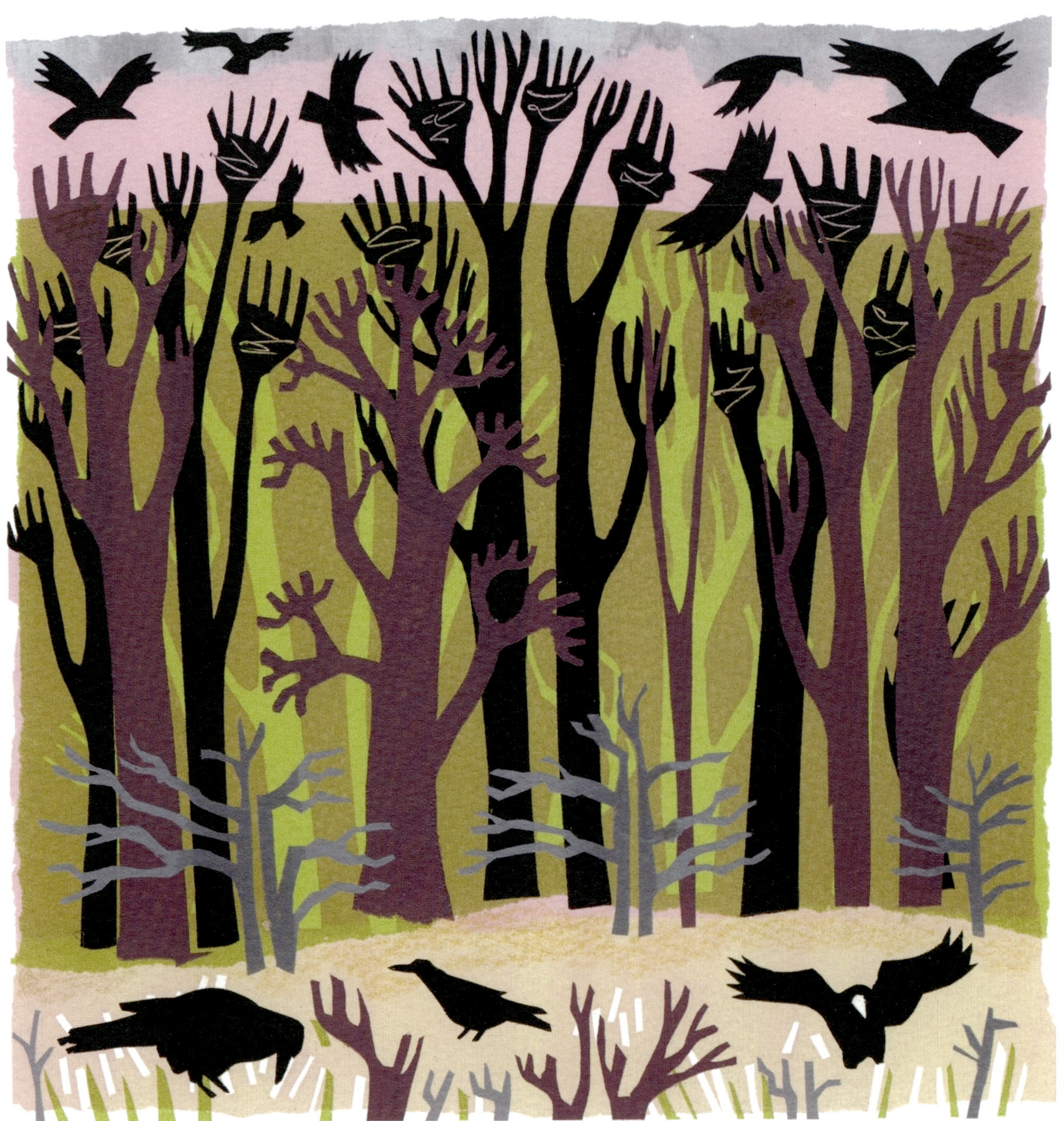

THE ROOK

The rook (*Corvus frugelicus*) is spring's noisiest messenger: 'This is partly bound up with the spectacle and symbolism of the rookery, which is as closely tied to our sense of spring as the sight of snowdrops, newborn lambs or, indeed, the sound of Easter bells.' Mark Cocker (*Birds Britannica*).

Over the land freckled with snow half-thawed
The speculating rooks at their nests cawed
And saw from elm-tops, delicate as flower of grass,
What we below could not see, Winter pass.
EDWARD THOMAS, 'Thaw'

A house with a rookery attached is blessed: Braal Castle's (Caithness) dates from at least 1775.

Oh! The merriest bird the woods e'er saw,
Is the sable Rook with his loud caw-caw.
ELIZA COOK, from 'The Rook Sits High'

Conversely, a house is cursed if the rooks leave. Maybe it says something that the last London rookery ceased in that dark historical year, 1916.

Rooks have 30 calls, but the caw or 'craa' (Scotland) predominates. In 'Maud' Tennyson is said to have written the line *Maud, Maud, Maud, Maud* in imitation. Pioneer eighteenth-century naturalist Gilbert White noted: 'Rooks, in the breeding season, attempt sometimes in the gaiety of their hearts to sing, but with no great success' (*The Natural History of Selborne*). They can sound like suddenly broken-voiced choirboys. If spring and rookery are indivisible, so are autumn, winter and the clamorous flight of rooks to woodland roosts.

More than 100 nests is an exceptional rookery. The record roost is 65,000 at Hatton Castle, Ayrshire. Older rooks have a building routine different from first-timers: one guards the nest; young couples go off to find sticks together and find those gathered stolen as a result. Rook pie or casserole, made from just-fledged birds perched on the verge of flight, is still legal but not the seasonal culinary event it was in living memory. May 12 was Rook-shooting Day, with boys and 'young ladies' sallying forth with low-calibre rook rifles. Rookeries are abandoned after the last nestlings fledge in June, and new roosts are found in woods into which the birds often whiffle (tumble aerobatically), frequently accompanied by jackdaws. Mark Hedges, editor of *Country Life* (2019) wrote: 'It's one of the greatest, most spectacular and under-rated wildlife sights in Britain.'

Following a decline, rooks have successfully adapted to modern farming and prospered. Britain boasts 40 per cent of the European population. Juveniles are hairy-beaked like carrion crows; adults have the dandruff-white skin patch where beak meets head. Having skin rather than clog-encouraging feathers is more efficient for earth digging. Long thigh-feathers give a knickerbocker look, and their plumage flashes sunlight, unlike other crows. They can be nest-robbers, carrion eaters, bird-feeder scavengers, even fishers, but plough is their preference, and leatherjackets (daddy-long-legs' grubs) a favourite food.

They excel as pets. 'Every home should have a rook, ' wrote Helen Mason, from the Isle of Man, to *Country Life* (27/7/16) of 'Kimi Rookinen'. In addition to grooming Phynn, the dog, Kimi weeds cracks, cleans gutters and, 'if he likes you', grooms humans too. He is not caged and divides his time between the household and rookery. Helen thought a dog was man's best friend, 'but Kimi is right up there'.

Since 2018, at France's Puy du Fou park, six rooks have been employed to pick up litter: a special litter box delivers a food morsel as a reward each time a bird deposits a cigarette end or other rubbish. No wonder a rook won a TV 'Bird Brain of Britain' competition.

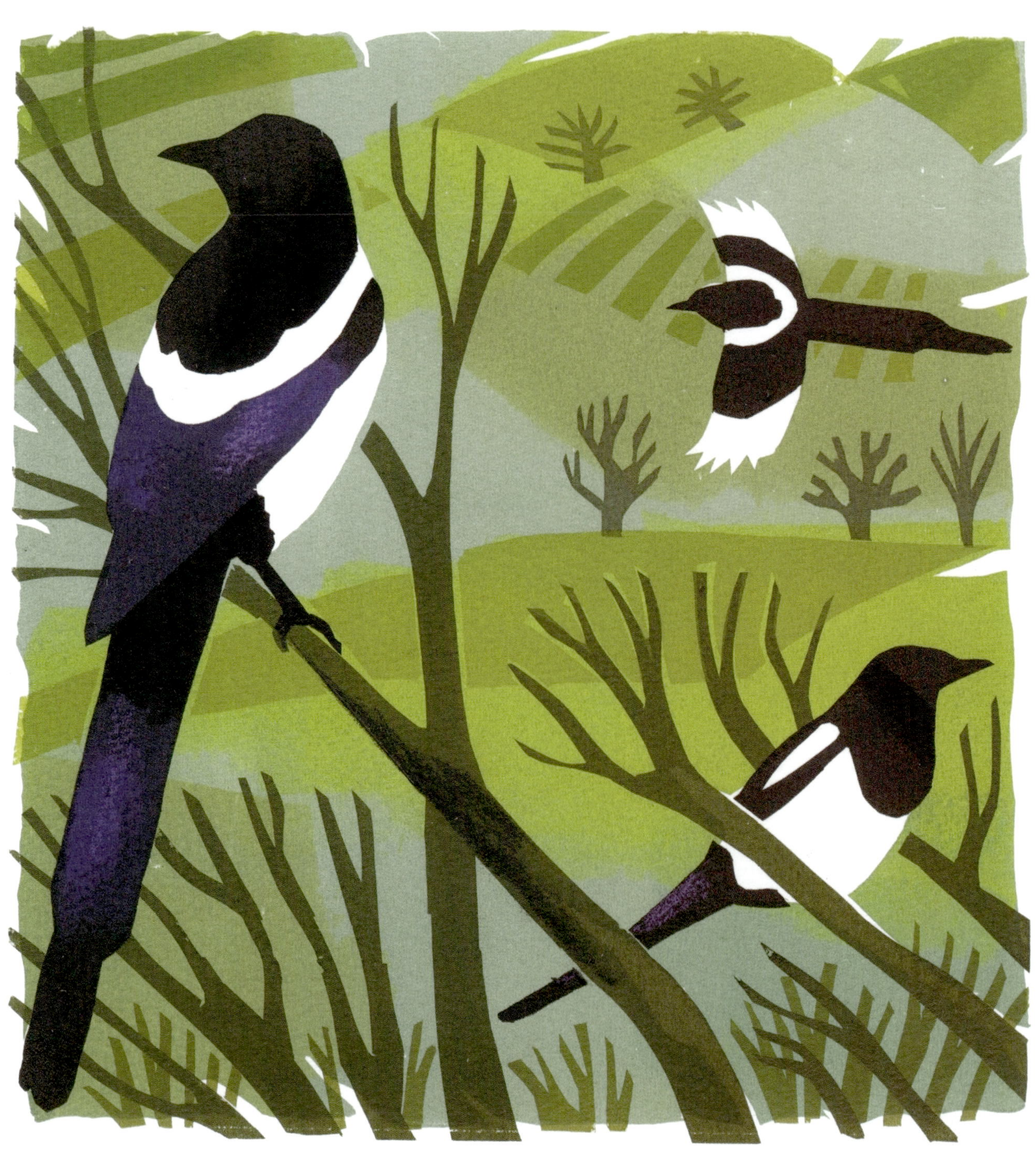

THE MAGPIE

I recently witnessed a sudden avian drama on a London pavement. A chattering magpie (*Pica pica*) was trying to kill a fledgling starling. My arrival sent the magpie flying. Then a cat shot in to finish the job. It had to be chased off before the starling could be caught and secreted at the heart of the thickest available bush. Michaela Strachan's advice on *Springwatch*, that empathy with the prey must not demonise the predator, went out the window. How much easier and longer songbirds' lives would be without cats and magpies.

Both have long attracted superstition, the magpie most fancifully. 'Pied' – i.e. 'mixed', usually black and white – meant it was cursed with a drop of blood from Satan, prince of darkness. It was also said to have refused shelter in Noah's ark, preferring to chatter and swear as the world drowned. We still salute the magpie, singly or grouped, to bring good luck or ward off bad. So the old rhyme persists in its many variations, the insistent 'sorrow' because magpies are monogamous:

> *One for sorrow, two for joy,*
> *Three for a girl, four for a boy.*

Or

> *One for sorrow, two for mirth,*
> *Three for a funeral, four for a birth.*

Or from Ireland:

> *One for sorrow, two for joy,*
> *Three for a kiss, four for a boy,*
> *Five for silver, six for gold,*
> *Seven for a secret never told.*
> *Eight for a letter from over the sea,*
> *Nine for a lover as true as can be.*

Outdoors writer David Profumo writes: 'In Dresden, magpie broth cured epilepsy, although in the Tyrol, it brought on madness. The Swiss swore it healed corns, if you chanted: *Aigi zigi, Aegest, I ha dreu Auga, ond Du gad zwa, ha, ha.*' (*Country Life*, 18/3/18).

There are now a million-odd magpies in the British Isles – the Irish blame their introduction on Cromwell – which often means their 'mischiefs' (groups) outnumber any ditty. One reason is lack of gamekeepers, who once controlled them to near extinction. A pair was not seen in Inner London till October 1960. Now they are everywhere.

March is when they nest build, often occupying abandoned dreys. These they turn into heaps the size of a small bonfire, the female sitting at the heart of it, her tail-tip sometimes visible. Their beauty is only rivalled among crows by the jay, the glossy black head offset by mantle, wings and tail refracting lustrous blue, emerald, bronze, pink and purple. The magpie was the bird of joy for China's last imperial (Manchu) dynasty, and the god of an ancient cult in Poitou. It has a crow's intelligence. Ovid noted its gift for mimicry; science has proved its rare capability of mirror recognition. Puccini's opera *The Thieving Magpie* has immortalised its reputation for stealing, even car keys found in a nest. 'Mag', an abbreviation of Margaret, is slang for a chatterbox; but their chattering might also be a death rattle.

> *The Magpies in Picardy*
> *Are more than I can tell.*
> *They flicker down the dusty roads*
> *And cast a magic spell*
> *On the men who march through Picardy,*
> *Through Picardy to hell.*
>
> T. P. CAMERON WILSON, from 'Magpies in Picardy'

Cameron Wilson was killed in 1918 at Hermies in France during the great German assault. He is commemorated on the Arras Memorial to the Missing and on the lychgate at Little Eaton church, Derbyshire.

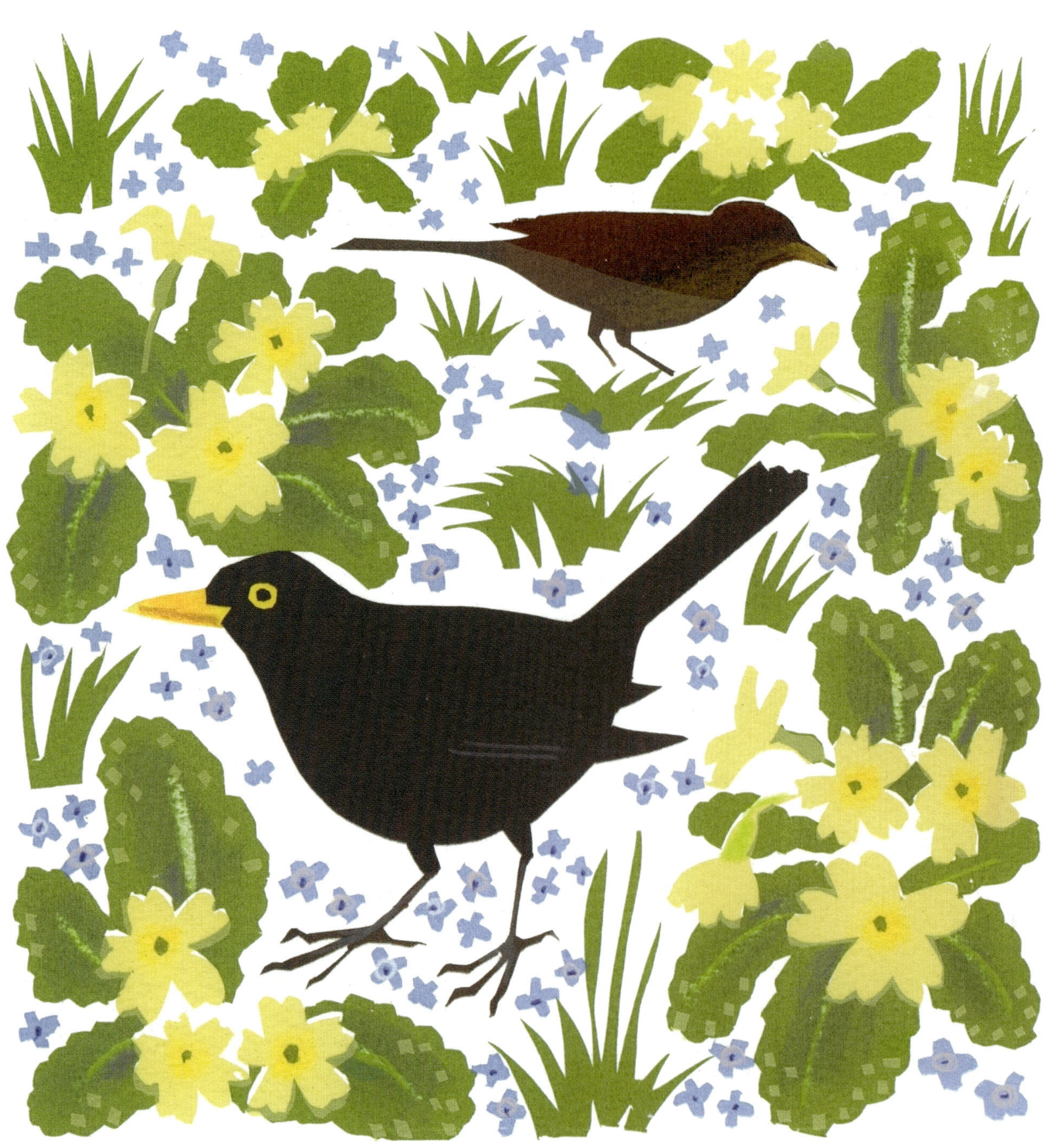

THE BLACKBIRD

Spring brings the song of the male blackbird (*Turdus merula*). The song may be a territorial defence but, as the diarist Alan Clark wrote of a 4am singer (15 June 1988): 'So clear and beautiful, as he went through his whole repertoire, he passed to me a lovely message of Nature's strength. Her powers of continuity and renewal.' (*The Alan Clark Diaries*). The song features most famously in Paul McCartney's 'Blackbird'. Denys Watkins-Pitchford (writing pseudonym 'BB') (1905–90), last of the great sporting naturalists – wildfowler, fisherman, artist, author, conservationist – wanted a recording of a singing blackbird at his funeral, which was sadly unfulfilled.

The pianist Joan Hall-Craggs made a study of a blackbird's song in her garden. It began with 26 basic 'musical' phrases repeated over several weeks. Then motifs were extended consistently – the same two phrases opening the song and a different two closing it. Over time the song had a sense of beginning and end, shape and balance. 'It does seem remarkable' she concluded, that the bird should effect the transition from repetition to a mature song 'by manipulation of those component basic phrases' (James Le Fanu, 'Profitable Wonders', *The Oldie*, March 2016).

The blackbird's urgent 'pink-pink-pinking' when mobbing predators or settling in winter roosts is another memory-stirring call, as is the liquid *sotto voce* 'pock' that punctuates the silence of a snow-filled wood.

That dark sound echoes the colly (coal) black male's plumage, hence the original 'four colly [not 'calling'] birds' in the carol. 'BB' wrote: 'The black of his plumage is…blacker than a mole's coat. It is the most beautiful black I know. Just one relief – and what a touch of genius it is – that bright golden bill' (*BB's Birds*).

The ousel-cock, so black of hue
With orange-tawny beak

sings Shakespeare's Bottom in *A Midsummer-Night's Dream*

'Ousel' variants were the convention until 'blackbird' took precedence from the seventeenth century. In Scotland, French merle is still *merl*, the auld alliance lingering in the name.

After a late-twentieth-century decline, British blackbirds have increased by over a quarter to come fourth in the UK's 2018 RSPB Big Garden Birdwatch count. It is native to Europe, Asia and North Africa and has been introduced to North and South America, Australia, New Zealand and even the Falkland Islands.

They can guzzle tadpoles and eat fish. The art historian Wendy Baron's resident Camden Town bird sat opposite as she filleted a fried trout: 'He darted onto my plate and carried the bone away to the ground. He could not open his beak properly because it was gummed up by the butter!' Eventually, all was well, and the bone was duly picked clean at her feet.

They can be equally versatile nesters. Fifty per cent die during the nesting season. I found a dead hedgehog in a nest high in a Northamptonshire yew hedge. A surfeit of nestlings had done for it. What a musical loss.

Ov all the birds upon the wing
Between the zunny show'rs o' spring…
The blackbird, whisslen in among
The boughs, do zing the gayest zong.
WILLIAM BARNES, from 'The Blackbird'

As Tennyson wrote in 'Early Spring':

The blackbirds have their wills,
The poets too.

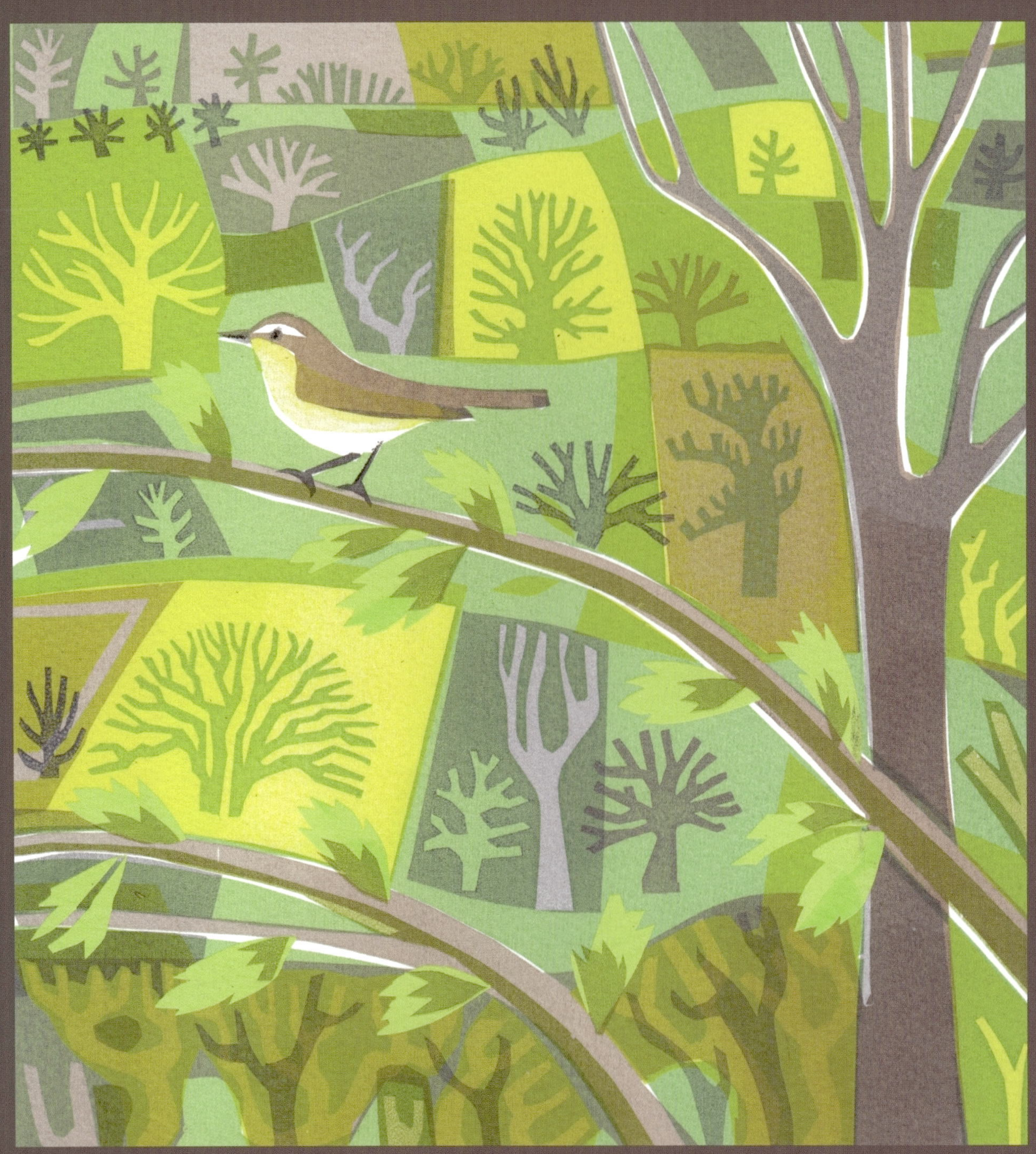

THE CHIFFCHAFF

No British bird declares the arrival of spring so recognisably and insistently as the chiffchaff (*Phylloscopus collybita*). *Pyllon* is Greek for 'leaf' and accurately describes its appearance as one of the 'leaf' coloured warblers, along with the willow and wood warbler. Even if you are lucky enough to see one in bare February, when the advance birds of its March army arrive, it is the call – 'song' too grand a word – that alerts one as decidedly as an alarm clock.

The doyen of twenty-first-century English bird writers, Derwent May, dared to render this as 'tink, tank, tink, tank' and was roundly rebuked by one of his *Times* readers for offering such an unromantic alternative. Of course he was right. The rhythm was exact and there is a hardness not conveyed by 'chiff chaff'. How difficult it is to convey the actual sound is illustrated by the onomatopoeic attempts in other languages: Dutch *tjitjaf*, German *zilpzap*, Welsh *siff-saff*.

The chiffchaff, often arriving in wintry conditions, is considered the hardiest of the warblers, although its almost identical cousin the willow warbler, whose poignant cadence is considered to be among the most beautiful bird songs, is more associated today with western and upland Britain. The hardiness of the chiffchaff – which principally winters around the Mediterranean, not in sub-Saharan tropical Africa like the willow warbler, means that, since the 1980s, there has been a steady increase to over a thousand remaining in Britain, even as far north as Shetland, throughout the year. These wintering birds favour habitats where insects are more abundant – wetlands or close to waste-water treatment works. The willow warbler remains more numerous at over 2 million summer migrants. The chiffchaff matches the blackcap at over a million.

Chiffchaffs nest low down in scrub and sometimes on virtually bare ground, as John Clare found to his astonishment and registered in his poem 'The Pettichaps Nest'; 'pettichap' then a common name for the bird in rural England.

Its nest – close by the rut-gulled wagon-road
And on the almost bare foot-trodden ground
With scarce a clump of grass to keep it warm!

Had he not put up the bird he would never have noticed the nest.

Built like an oven through a little hole
Scarcely admitting e'en two fingers in,
Hard to discern – the birds snug entrance win.
'Tis lined with feathers warm as silken stole,
Softer than seats of down for painless ease,
And full of eggs scarce bigger even than peas!
Here's one most delicate, with spots as small
As dust and of a faint and pinky red...

Well! I declare it is the pettichap!
Not bigger than the wren, and seldom seen,
I've often found her nest in chance's way,
When I in pathless woods did idly roam;
But never did I dream until today
A spot like this would be her chosen home

There the 11.1 million cats in the UK, which kill 55 million birds annually. The RSPB assures bird lovers there is no scientific evidence this has been an important factor in the decline of UK birds – 25 per cent (40 million) since the 1960s. In the great scheme of things, this is doubtless true, as Prince Philip, Sir David Attenborough and countless other authorities have long warned, the relentless rise in the human population is the prime cause, of too many UK cats as of everything else. But cats' easiest prey are nestlings and fledglings, which makes ground and low-scrub nesters like the chiffchaff particularly vulnerable. Cat owners should at least bell their pets to give birds some warning.

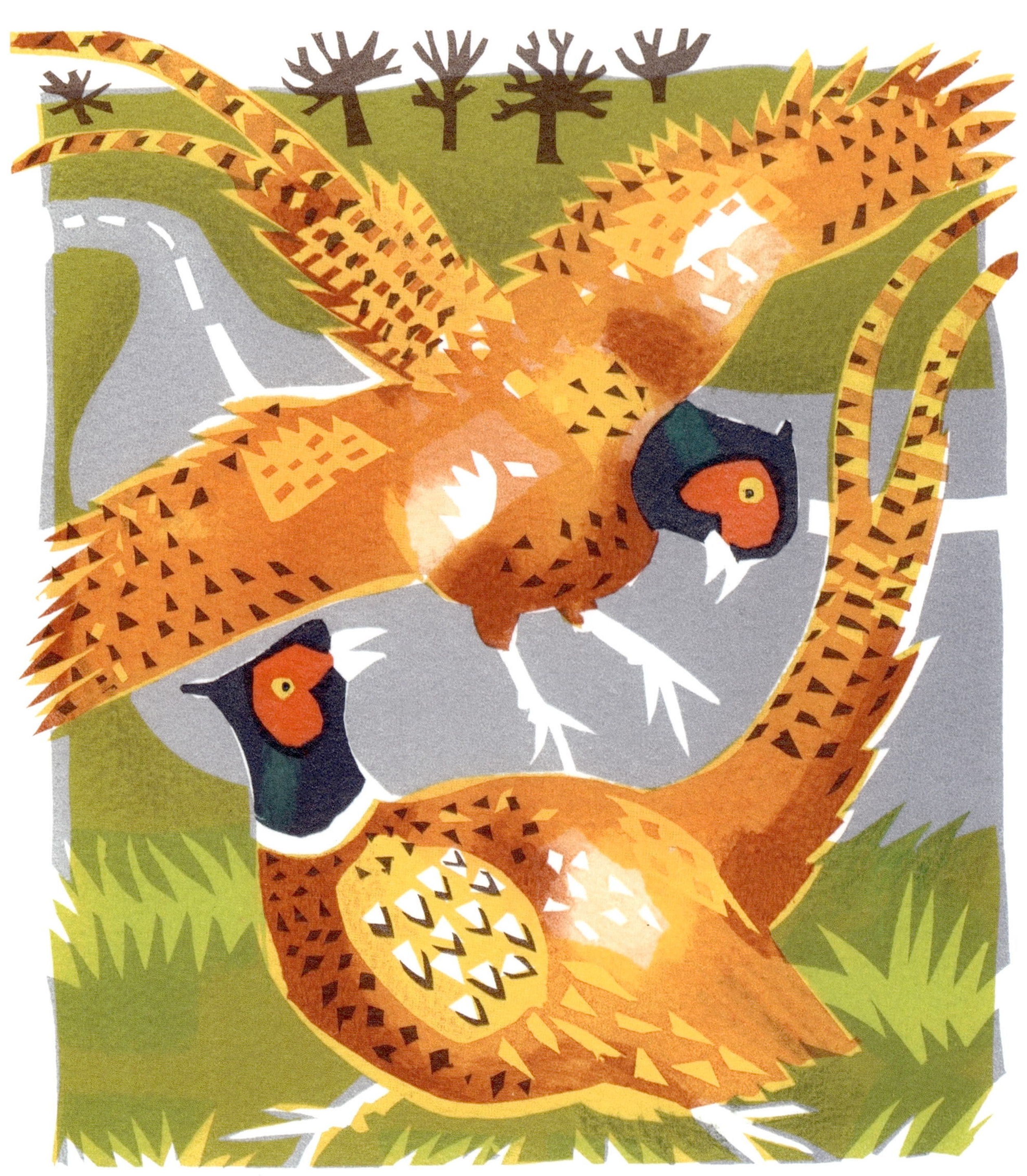

THE PHEASANT

Cock stubble–searching pheasant, delicate
Stepper, Cathayan bird, you fire
The landscape, as across the hollow lyre
Quick fingers burn the moment: call your mate
From the deep woods tonight, for your surprised
Metallic summons answers me like wire
Thrilling with messages, and I cannot wait
To catch its evening import, half–surmised.
Others may speak these things, but you alone
Fear never noise, make the damp thickets ring
With your assertions, set the afternoon
Alight with coloured pride. Your image glows
At autumn's centre — bright, unquestioning
Exotic bird, haunter of autumn hedgerows.

SIDNEY KEYES, from 'Pheasant'

The common pheasant (*Phasianus colchicus*) is worthy of the name Colchicus (modern Georgia), home of the fabled golden fleece of Greek legend, goal of Jason and his Argonauts. The cocks of its 30 subspecies range in colour from green-black to honey with pale blue rumps; their pelts superior to any fleece, however golden.

Autumn matches their plumage, but spring is the cock's glory time: head tufts prominent, scarlet skin patch swelling to crest above its eye. That is when they scrap and proclaim their domination with a 'cock cocking' crow and wing flurry. I am indebted to Matt Ridley, world authority on the bird, for the following: 'Pheasants are among the five per cent of birds that do not have two parents rearing the offspring. They are socially polygamous – one male with several females. Not having fathers involved in feeding chicks has freed females to choose fathers for their offspring by some other criterion than "is he a good provider" and "is he unmarried?" So females choose males on the basis of gorgeous plumage and their ability to dominate other males, indications of good genes to some extent. Hence they cluster with certain males.'

Pheasants have been forever hunted for food, and it was Prince Albert who brought the driven-pheasant *battue* to England. 'Arrows at Agincourt!' replied the late Bobby Corbett, legendary sportsman and wit, when asked to describe an abundance of driven birds. Technology has rendered some pheasant shooting absurd. A clueless Chinese shooting party arrived on Exmoor with newly purchased guns and wearing white snowsuits, designer jeans and baseball caps. Having got cold and wet, they each paid £1,000 for immediate delivery of traditional tweeds and footwear ('Super-rich pay to bag 2,000 kills', *The Times*, 19/1/19). The writer John Kampfner interviewed an Arab Prince. He was taken to 'a vibrant green shooting estate' in the desert where pheasants were flushed by beaters in plus fours: 'We could have been in Wiltshire except for the searing heat' (*The Rich: From Slaves to Super-Yachts*).

Shooting in the UK is so popular that in 2019, it is the biggest threat to fox hunting: 570,000 men and increasingly women have licenses. They mostly attend the small shoots familiar to 5 million BBC Radio 4 listeners of *The Archers*, the world's longest-running radio serial. Over half the UK's 5,000 pheasant shoots run at a loss; by contrast, commercial shoots (over 40 million birds reared annually) are big business. They help boost the annual economy by £2 billion and provide 75,000 jobs (see The Grouse, page 121), offsetting the decline in rural tourism and vegan hostility to meat production. Paul Langley of Cramers, north London's fabled butchers, has to remind mothers complaining of his hanging pheasants that he is not a sweet shop.

But commercialism is problematic. Of the 40 million pheasants reared, too few reach the table. Some shot birds are unacceptably trashed. The National Trust and Welsh Assembly are just two institutions that ban shooting. Nor does an 85 per cent urban population eat game, albeit its recommendation as an ultimate health-food resource.

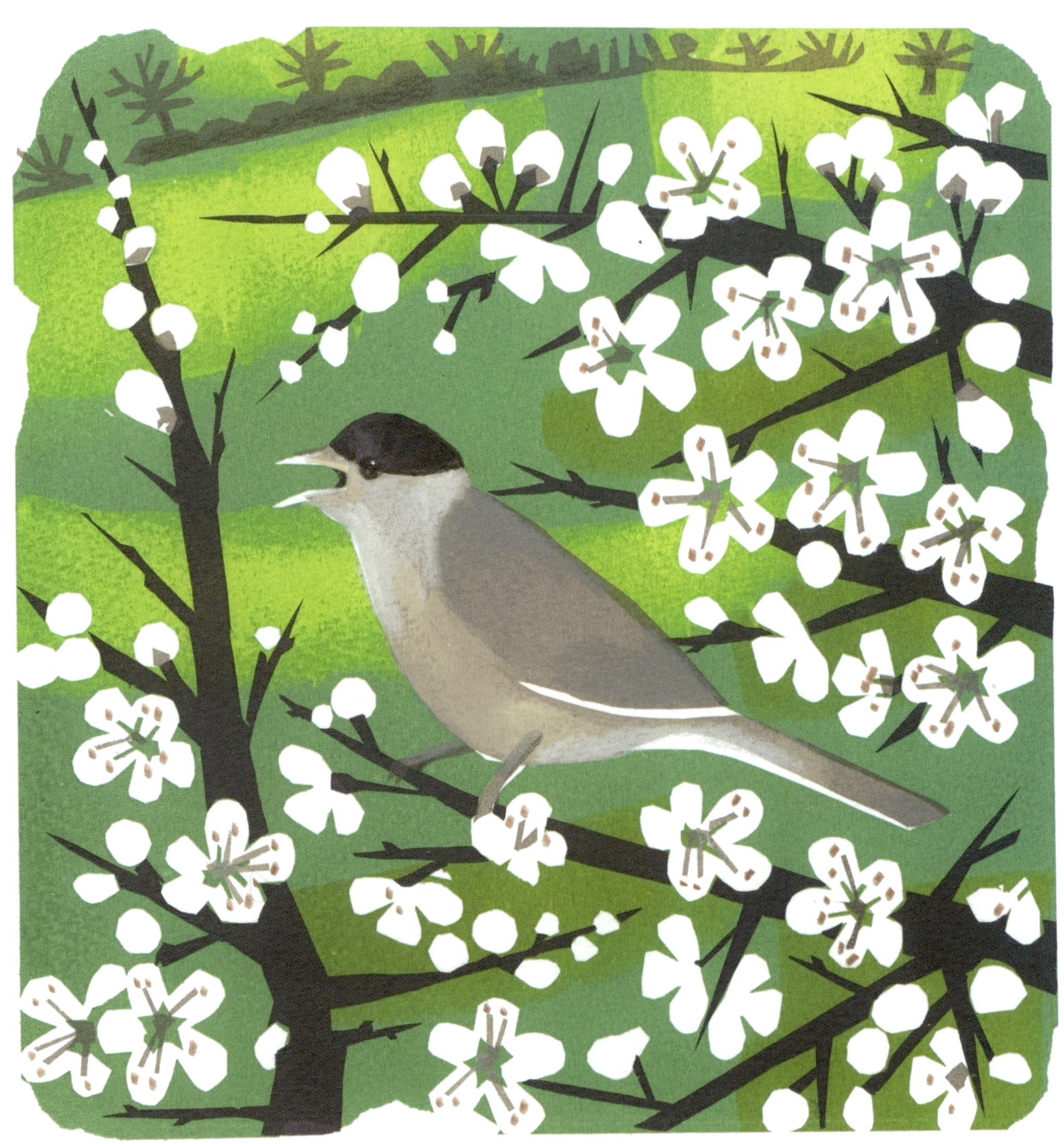

THE BLACKCAP

Warbler species can look so alike that song is the more reliable guide. Fortunately, because its numbers are rising, one whose appearance is as distinctive as its song is the blackcap (*Sylvia atricapilla*), thanks to its neat skull cap, black in the male, rust brown in the female. They are also easier to see because 3,000-odd, most from Germany, now winter here, not least due to the abundant bird food we provide. Unlike the garden warbler, the blackcap does not migrate from tropical Africa; but it remains a predominantly summer visitor – from Morocco and southern Europe. Dave Clifton, a bird-ringer, was amazed in 2014 to catch a blackcap in Portugal he had ringed in Staffordshire a few weeks before. London is a good indicator of its rise in numbers: from virtually none in the 1940s to 82 per cent by the mid-1990s. Wimbledon Common in 2018 hosted more than a hundred breeding pairs.

'Migratory birds are usually imagined by biologists to have an inborn programme that directs the migratory process on the basis of compass orientations derived from the sun, the stars and the magnetic sense,' writes the biologist Rupert Sheldrake. But how does this programming work when skies are overcast and magnetic poles wander, which they do? And how does it explain new migratory cycles like Britain's winter influx of blackcaps? 'According to the conventional genetic programming theory, the evolution of a new migratory pathway would depend on chance mutations that affect the genetic programming. Then these mutant genes would have to be favoured by natural selection over many generations for a new gene to emerge.' Sheldrake proposes that 'neither new destinations nor the migratory pathways themselves would be coded in the genes' but rather remembered by 'morphic resonance' and thus 'new migrational pathways could be established in a single generation'. He explains the concept of morphic resonance as: 'Each individual inherits a collective memory from past members of the species, and also contributes to the collective memory, affecting other members of the species in the future.' Migration remains largely unfathomed, and he is happy with his concept as a contributory explanation until science proves it right or wrong.

Trim and elegant like most warblers, blackcaps are always a joy to see, but it's the male's rambling song and vociferousness that causes most excitement among bird lovers. John Clare called it 'The March nightingale':

From 'The March Nightingale'

The blackcap is accordingly also called the 'northern nightingale'. Gilbert White described its song as 'a sweet, wild note', hence the 2015 book on birdsong, *A Sweet Wild Note* by Richard Smyth. The blackcap's song is Smyth's favourite: 'Cracked, drunken, loud, littered with chitters and whistles, and generally all over the shop – it doesn't sound like music to me, and I think that's sort of the point. The blackcap doesn't give a damn what it sounds like to me. Birds never do.'

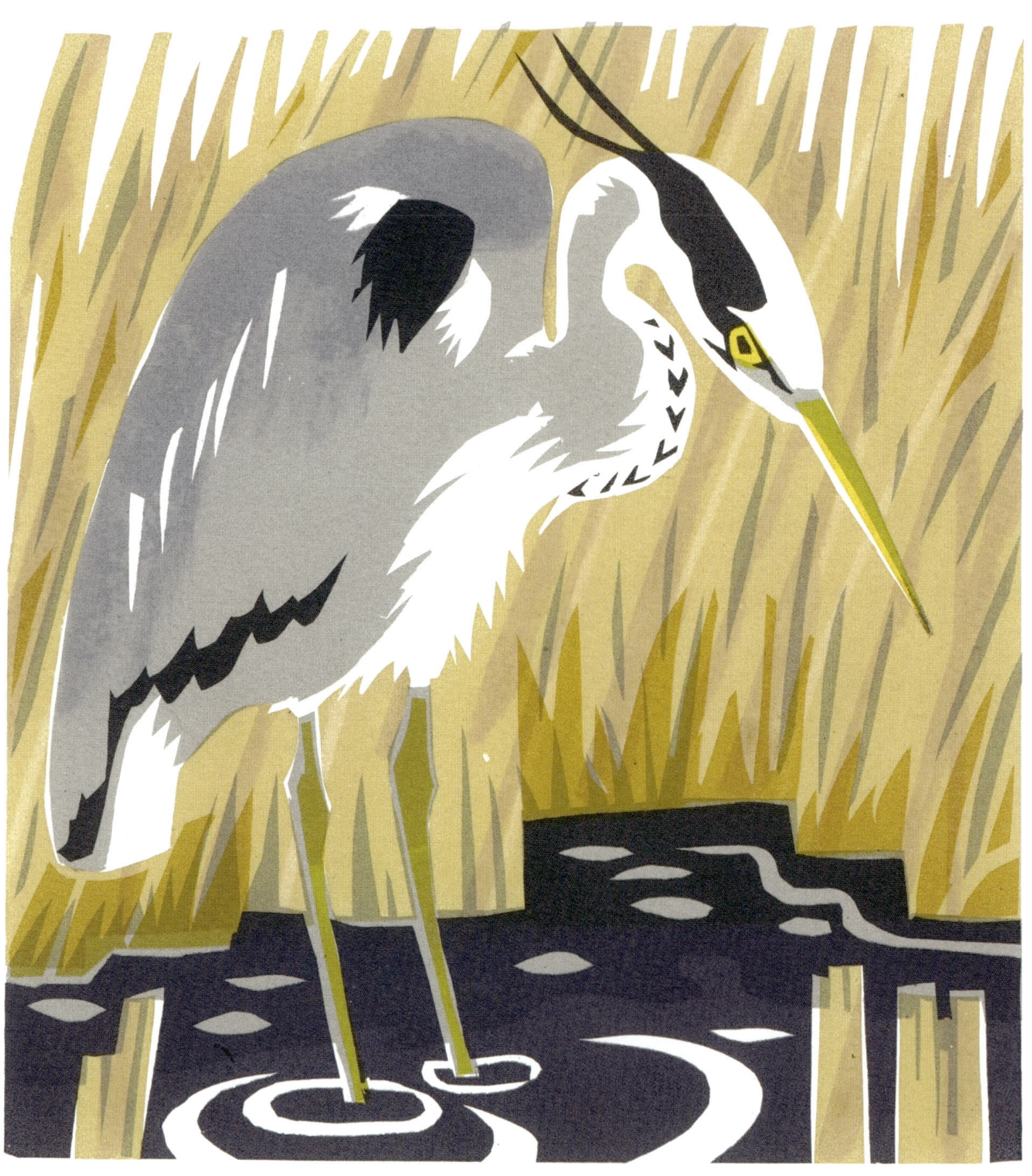

THE HERON

In Britain, 'heron' means the grey heron (*Ardea cinerea*) of which there are 12,000 breeding pairs, more than at any time since surveys began with the 1928 heron census (4,000 breeding pairs). The census, organised by *British Birds* magazine, has a unique historical place as the progenitor of all surveys, the method soon adopted by the social sciences. Key figures in this interwar field were originally ornithologists. Max Nicholson (see The House Sparrow, page 19), who organised the heron census, apart from becoming a towering figure in conservation, inspired the independent think tank, Political and Economic Planning, which helped lay the foundations of the NHS; and the charismatic Tom Harrison, an organiser of the 1931 great crested grebe inquiry, later founded and directed the pioneering social research organisation, Mass Observation.

There are hundreds of heronries, from the Isles of Scilly to Shetland. The best known is in Regent's Park, where the birds benefit from London Zoo with its daily handout of fish for the Humboldt penguins. They first nested in 1968; now it is the most central and fourth-largest heronry in London. Herons need mature trees to support the weight of their stick-piled nests. In treeless country, they resort to cliffs, islets and heather. The breeding season can stretch from March to September, sexes sharing incubation.

'The great number of riuers riuulets & plashes of water makes hernes & herneries to abound in these parts. Yong hensies being esteemed a festiuall dish & much desired by some palates,' wrote Sir Thomas Browne in the seventeenth century (*Notes and Letters of the Natural History of Norfolk*). Herons, called 'harnsa' in old East Anglian dialect, were popular prey for falconers. This explains the apparently nonsensical line in *Hamlet* 'When the wind is southerly, I know a hawk from a hand-saw' – a mistranslation, perhaps by the Warwickshire Bard himself, of 'harnsa'/heron. Herons are no longer eaten but still abound in Norfolk, and can only be legally killed if threatening fish stocks.

On 17 June 1990 Alan Clark, then a cabinet minister, wept bitterly after shooting one that was decimating his Saltwood Castle carp in Kent: 'I cussed and blubbed up in my bedroom…I was near a nervous breakdown. Yet if it had been a burglar or a vandal I wouldn't have given a toss. It's human beings that are the vermin.' (*The Alan Clark Diaries*)

A heron will prey on anything aquatic, from rats to substantial ducklings. A fish pond in the garden, however small, will catch its eye.

Old Nog — Henry Williamson's coining. Spearer of fat rats; eel-wrestler. Fun, giving it names, affectionately prefaced by "old": spearheid grunzel, highshanks, grey jack, langthrapple; for can one imagine it young, young at least in spirit? Or imagine it female, either? In my head they are always male, hodden-doun, long coat belted with baler twine; stubbly and jut-cheek-boned, bird of dreich weathers, distrustful to a fault: an expert at ignoring the question that anyway you couldn't get close enough to ask. Suspicious rifler of the pockets of any query, especially hackled at 'Where you fae?' And so, all the more intriguing. A colony near Beith, in a stand of Scots pine in the middle of a field. You would never have sussed till you walked towards it and one by one, big grey winged, they erupted lugubriously from the tangled needles and grounded at distance on the heights around, each keeping an acute eye out. Like meanings from arcane poems.

GERRY CAMBRIDGE, 'Ardea cinerea'

Aesop's fable told of the heron removing a bone from the wolf's throat for a promised fee. The wolf said not biting off its head was payment enough. Once a cad, always a cad.

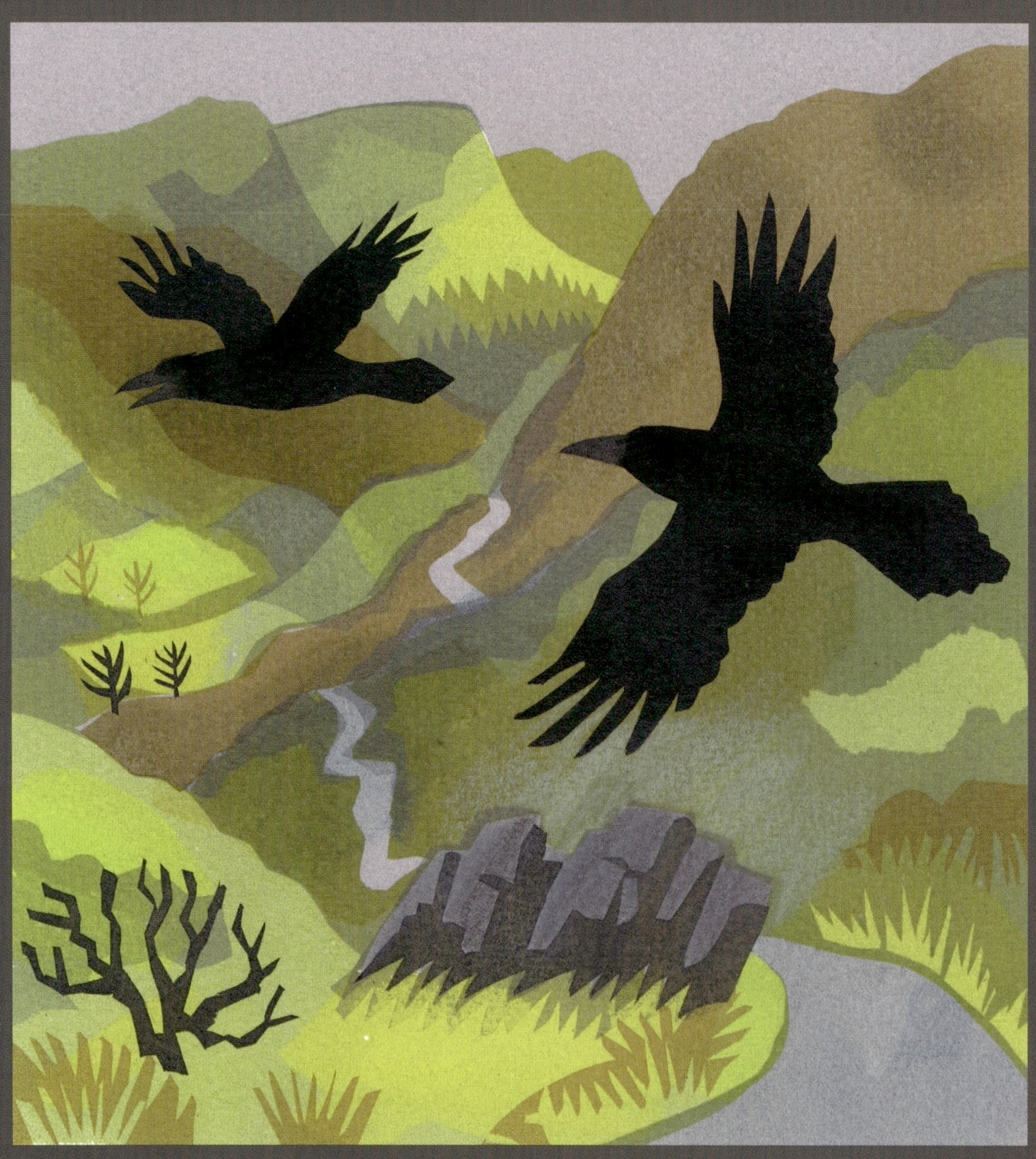

THE RAVEN

The raven, king of crows, is a legendarily mystical bird. In *The White Goddess*, Robert Graves cites Bran, the Celtic Crow-god, as the equivalent of Aesculapius, Apollo, Cronos and Saturn. Worship of the healing Aesculapius was combined with worship of the Thundergod, the bull or ram, also known as Jupiter, Tantalus, Telamen and Zeus. 'Bertram' means 'bright raven'.

It is suitable a Welshman should proclaim the mythic reach of the raven (*Corvus corax*), king of crows (5ft/1.5m wingspan, diamond-shaped tail) because central Wales has more than probably anywhere in the world; its bottomless bass 'kronk' is the voice of the Welsh mountains.

The scavenging of the raven, 'corbie' in Scotland – also applied to carrion and hooded crows – has made it mankind's familiar for aeons, its doom-laden reputation inseparable, 'in the estimation of the vulgar and superstitious' (*Jamieson's Scottish Dictionary*), from gibbets and battlefields:

> *As I was walking all alane,*
> *I heard twa corbies making a mane;*
> *The tane unto the 't'other say,*
> *'Where sall we gang and dine to-day?'*
>
> *In behint yon auld fail dyke,*
> *I wot there lies a new slain Knight;*
> *And naebody kens that he lies there,*
> *But his hawk, his hound and lady fair."*
>
> ANON, from 'The Twa Corbies'

'Ravenous' is well earned. Charles Dickens had two pet ravens, one tore up and swallowed the greater part of a wooden staircase. His favourite became 'brave Grip', Barnaby's companion in *Barnaby Rudge*. It inspired Edgar Allan Poe's *The Raven*, which made Poe famous. Stuffed Grip is the pride of the Philadelphia Free Library's Rare Book Department.

Ravens range from deserts to tundra and up to 21,000 feet (6,400m) on Everest. They are intrinsic to cultures west and east, from North American totem poles to Viking banners. Noah's raven, which failed to return to the Ark, preceded the dove and is possibly the first named bird in literature. In the New Testament they come before the lilies of the field: 'Consider the ravens: for they neither sow nor reap; which neither have storehouse nor barn; and God feedeth them: how much more are ye better than the fowls?' (Luke 12: 24). Their former ubiquity in the UK is illustrated by 400 'raven' place names (60 in London). They have increased by 45 per cent since 1995, and in 2019 Britain's 12,000 breeding pairs are reclaiming the eastern lowlands vacated since the nineteenth century. Reclamation receives growing opposition from Wiltshire to Caithness. Ravens kill newborn lambs in particular, tearing out tongues and pecking out eyes, although all 'hoodies' do this.

Most famous are those that guard the Tower of London. The belief that if the ravens leave the Tower the Crown and Britain will fall was a Victorian invention. Churchill nonetheless ordered the Blitz-diminished flock to be raised to full strength (six, plus a reserve) in 1940. Under Attlee's post-war Labour government the raven garrison was briefly zero. The post of Raven Master dates from 1968. The ravens have military ID cards and are dismissible for bad conduct. In 1986 Raven George was posted to the Welsh Mountain Zoo for destroying TV aerials. In 2019 the Tower garrison is at full strength. The mischievous Merlina is the public's favourite, a TV celebrity with a following on Twitter, Instagram and Facebook. The longest-lived Tower raven was 44.

Ravens are famously aerobatic. Roc Sandford, owner farmer of the Scottish island Gometra, writes of a walk: 'A solitary raven accompanied me, repeatedly flying upside down a few feet above my head, and making a "prooking" and low "pinging" sound. I thought it was showing off, but a friend suggested it was courting me.'

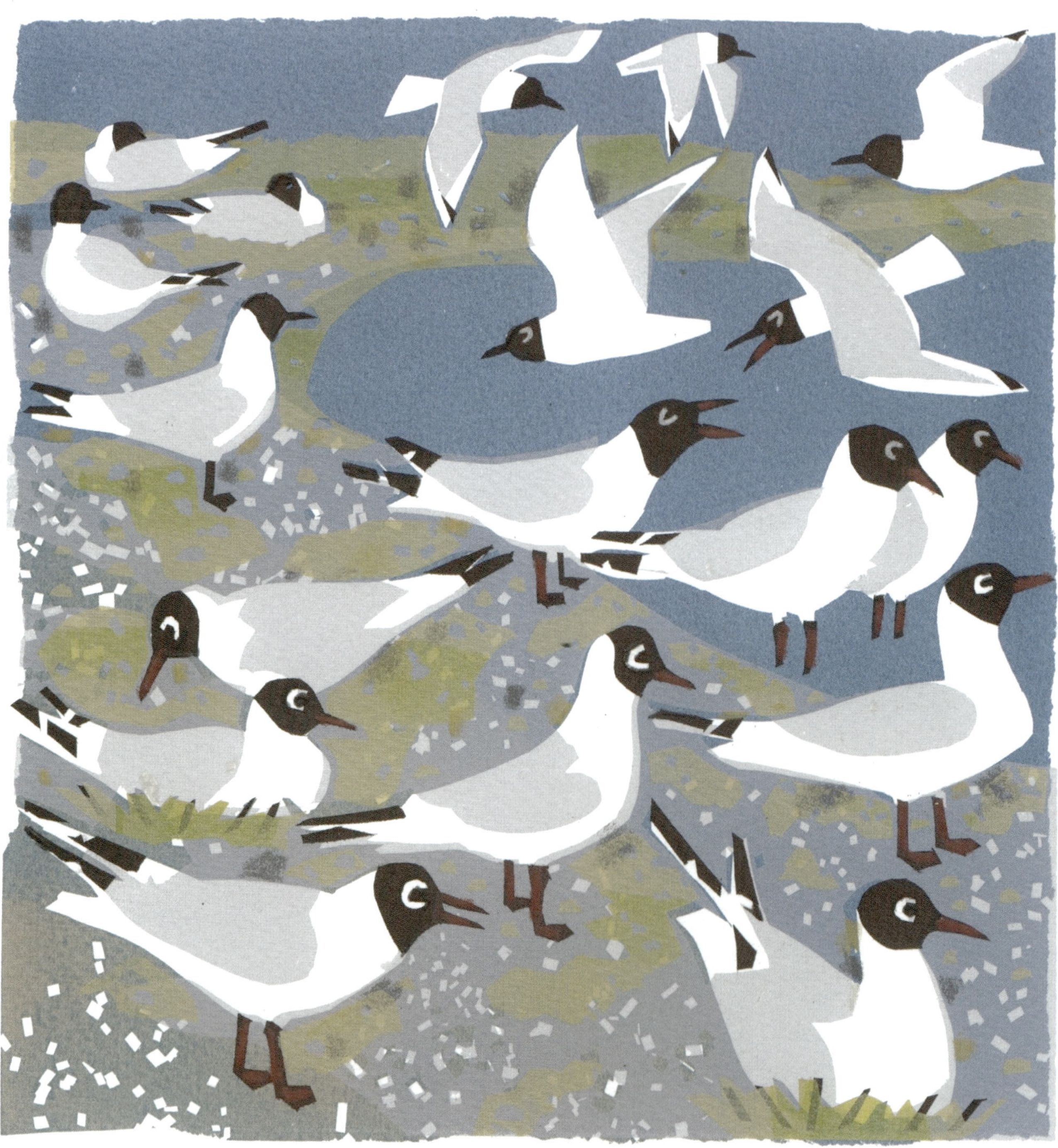

THE BLACK-HEADED GULL

Robin Douglas-Home – eldest son of Henry – classics scholar, soldier, pianist, copywriter of 'Have a break, have a Kit-Kat' – wrote a novel *Hot for Certainties* (1964). The main character, a young officer, finishes the book parked in the rain on the Mall, sobbing drunken tears of anger and self-pity. Then he grabs 'a cardboard box full of gulls' eggs', destined for his mother's wedding breakfast, walks unsteadily across the road and climbs the steps of the Victoria Memorial. 'He picked out a gull's egg from the box and flung it at the unsmiling head…Another and another he flung, bawling out curses as the eggs cracked and splashed on the impervious white-marble figure.' A policeman appears with a torch. 'Gulls' eggs, eh? You must be well lined to be throwing these about.' After the hero's identity is established he is let off with a deferential caution. End of novel.

'Gulls' eggs' are those of the black-headed gull (*Chroicocephalus ridibundus*), for epicures an annual highlight. In the UK they are the only wild-bird eggs it is legal to collect (those of pheasants and quails are farmed). The few licence holders must log their take: 27,842 eggs were sold in the 2015 season (1 April–15 May), each costing up to £10. It contrasts with our annual consumption of 13 million hens' eggs. Gulls' eggs are a treat indeed: beautiful, with their pointed, freckled, jade shells, and easy to peel; the whites faintly blue and translucent like mother-of-pearl, yolks pink, and tasting pleasantly, if slightly, of fresh-mown grass. They are usually sold hard boiled. Douglas-Home was right: order them uncooked. Birders are happy with their consumption – b-headeds nest early and can occupy the empty sites of migrant terns. Gulls are predatory, so they also eat other species' eggs and nestlings.

They are the most ubiquitous and land-dependent of gulls (*Laridae*), the ones dominant on playing fields and in the wake of ploughing tractors. The black, actually dark-brown, hood is breeding plumage and applies only from late spring to early autumn. Most of the year the head is white with a 'beauty' spot – varying degrees of smudge appearing as summer and winter approach. Juveniles do not acquire a hood until they are two years old. Winter-plumaged black-headeds are the ones most of us see since winter is when the 130,000 resident breeding pairs, which disappear to their bogs and wetlands to nest, recolonise city parks and lakes; their numbers boosted by 2 million winter migrants from mainland Europe.

It is then the fun starts, as no bird is so enjoyable to feed. Almost a century ago Lord Grey wrote: 'In London, black-headed gulls, that are entirely wild birds, come in numbers in autumn and winter, and will feed, or at any rate snatch food from the hand.' The best game is to throw food in the air to test their astonishing aerial agility. No wonder they can also hawk for insects. The 'laughing' (*ridibundus* in Latin) and carping ('Kehaar' the black-headed's name in *Watership Down*) birds become as excited as the throwers, the frenzy attracting others from far and wide.

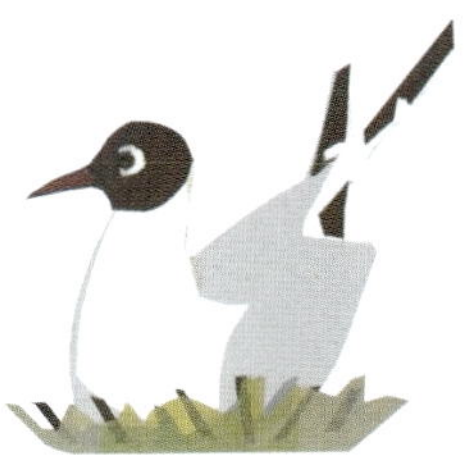

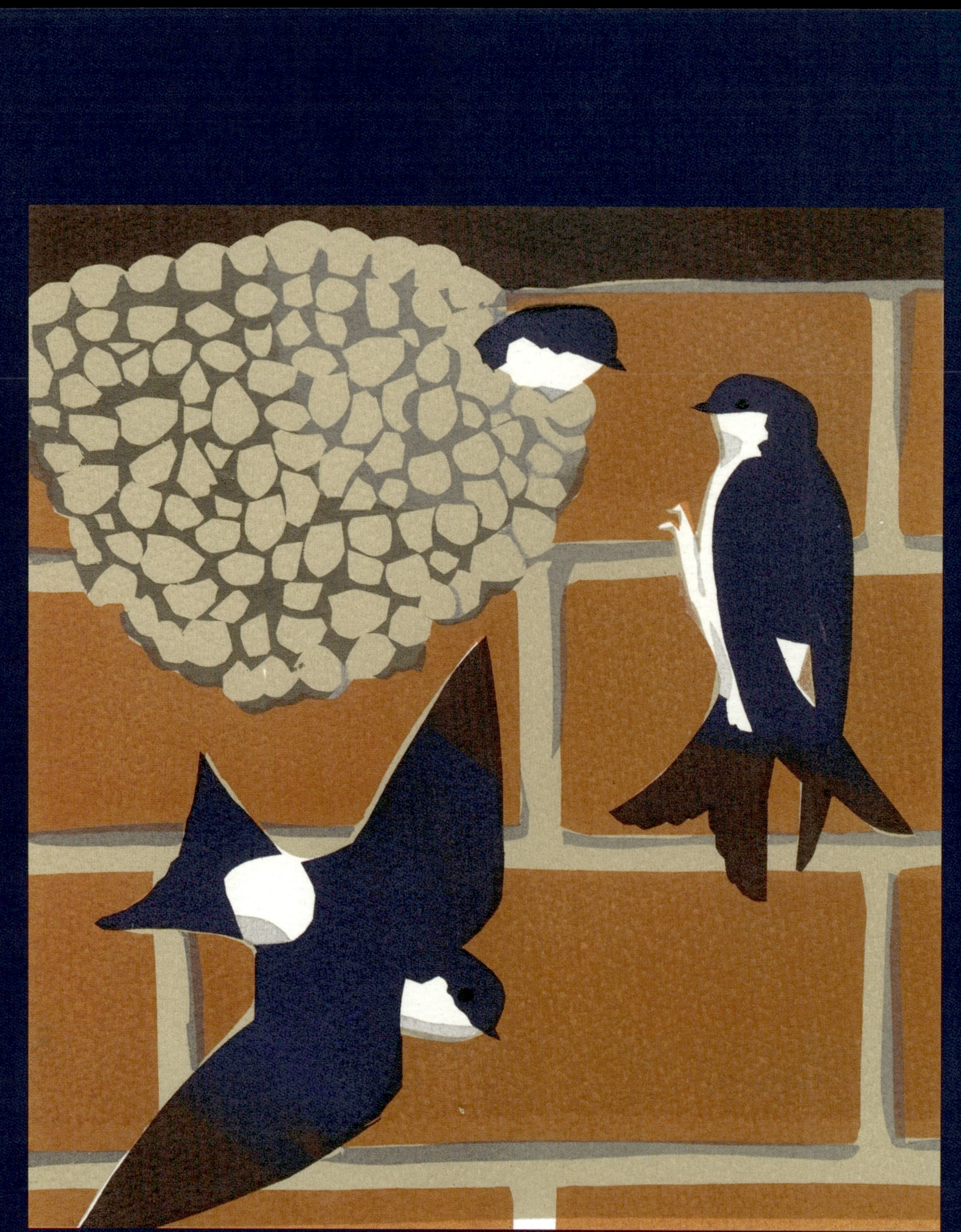

THE HOUSE MARTIN

In his *Oxford Dictionary of British Bird Names*, W. B. Lockwood describes the martin – which usually means the house martin (*Delichon urbica*) rather than the brown, bank-nesting, sand martin – as 'the oldest known example, by far, of the use of a Christian name to denote a bird'. Perhaps its cloaked appearance earned it the name; or that it is first seen here in March and last in November: March deriving from Roman Mars, god of war, hence martial St Martin (from Martius), early Christian martyr, whose feast day is Martinmas, 11 November. St Martin, when a soldier, compassionately gave his cloak to a beggar. He recognised the beggar as Christ, converted and entered the priesthood to be 'Christ's soldier'.

This guest of summer,
The temple–haunting martlet, does approve
By his loved mansionry , that the heaven's breath
Smells wooing here: no jutty, frieze,
Buttress, nor coign of advantage, but this bird
Hath made his pendent bed and procreant cradle:
Where they most breed and haunt, I have observed
The air is delicate.

Says Banquo in *Macbeth* (Act 1, Scene 6). John R. H. McEwen, writer, actor and Chair of the Duns Players, wrote to me: 'I don't think there can be any play so ornithological.'

March brings the swallow, whereas the chunkier, black-and-white house martin appears in numbers from April. Swallows will nest in living rooms (see The Swallow, page 113), so cannot be said to avoid houses, but in colder Britain, they have to be content with barns and garages; house martins prefer eaves and window corners. Both very occasionally overwinter in south-east England.

House martins were Gilbert White's favourite birds. He noted that on arrival they paid 'no attention to the business of nidification, but play and sport about'; only around 'the middle of May, if the weather be fine' do they start to nest, often using the crust from the previous year as a foundation. These mud cones, sometimes fused to another, have an echo of the traditional coned mud huts of tribal villages in the African rainforests where they spend their winters. The swallow's cup requires support, the house martin glues its cone to a vertical wall, using its stubby forked tail as a fulcrum while adding the mud, pellet by pellet. White noticed that by building in the morning and 'dedicating the rest of the day to food and amuse' they allowed the mud to set.

Why do martins like one house and not another, albeit identical? There is a semi-detached villa in London's Primrose Hill with deep eaves. It has a substantial if declining colony (eight nests in 2018), whereas its adjoining villa had one nest. No other villa has any. Indeed, is there another central London colony?

There is a worrying general decline, its cause not helped by the difficulty of tracking their African movements, which remain largely uncharted. Like swifts they stay high above the Congo forests. The latest estimate is a UK and Ireland migration of up to 600,000 breeding pairs. As with swallows and sand martins, Ireland now has the greatest abundance, and it is England – notably the south-east – that shows the greatest decline. Hirundinidae thrive where insects thrive, and Ireland still has abundant outdoor livestock.

Alone of the Hirundinidae the house martin has feathered feet. White also liked the way it twitters 'in a pretty, inward, soft manner' in the nest.

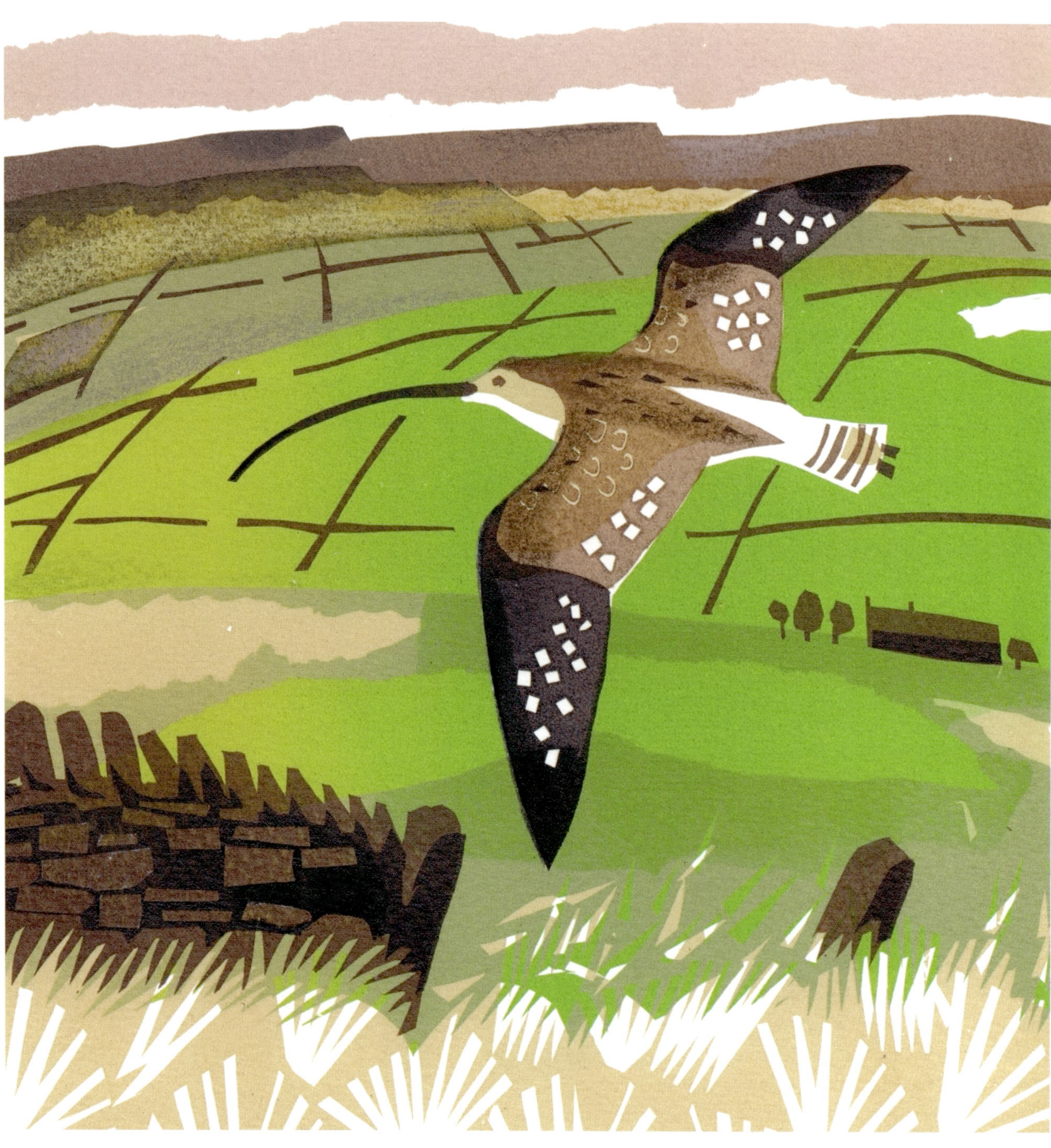

THE CURLEW

The curve-billed curlew (*Numenius arquata*, 'arced like a bow') is Britain's largest wader. The bill enables it to scoop its favourite lugworms whole. The tug of straight-billed waders can break the worm, thus losing the nutritious head.

The name is an accurate rendering of its haunting call, inseparable from upland wildernesses in the breeding season, when the bird comes inland from its winter flocking on the foreshore. The call comes to a bubbling crescendo during the male's aerial mating displays. Robert Burns never heard it 'without feeling an elevation of the soul'; and Lord Grey wrote: 'Of all bird songs…there is none that I would prefer to the spring notes of the curlew'. Whaup, also onomatopoeic, persists in Scotland. Burns's Galloway moors and mosses, once a refuge for persecuted Covenanters, were ideal whaup-nesting country; the two brought together in Robert Louis Stevenson's sonnet 'To S. R. Crockett':

Blows the wind today, and the sun and the rain are flying,
  Blows the wind on the moors today and now,
Where about the graves of the martyrs the whaups are crying,
  My heart remembers how!

The conifer carpets that now disfigure Galloway and so many previously bare uplands are a significant contribution to the bird's precipitate decline – until 1981 its numbers still allowed it to be legally shot as a sporting quarry in the UK. In Ireland, largely due to bog drainage, it has declined 90 per cent, and in Wales 80 per cent since 1990.

In 2016 the radio/TV producer and environmental writer Mary Colwell, alone and at her own expense, made a 'Curlew Walk' from the west of Ireland to the Lincolnshire coast during the breeding season, to raise awareness of the bird's plight. 'Curlews are in need of so much help,' she wrote. 'In one generation they have disappeared from many places and are particularly endangered in the west and south of the British Isles.' (*Curlew Moon*)

The bird's twin curses of forestation and intensive farming, including wind farms – nest numbers can be halved within 550 yards of a turbine – are further symptoms of the decline of the old country order, where the sporting estate, in the stewardship of Lord Grey and his nature-loving kind, ensured the vermin control which protected so many bird species. Forests are safe havens for predators. As *British Birds* magazine reported (November 2015): 'In the UK, Curlew population changes are inversely related to the area of woodland surrounding breeding sites, and positively related to gamekeeper densities.'

Legend attributes the notorious difficulty of finding its nest to a blessing of the Welsh St Beuno, whose book of sermons was saved by a curlew from the sea. But there is no defence against its modern man-made enemies. Mary Colwell concludes that the curlew of all UK birds should be given the highest conservation priority. More power to her and her like, otherwise we may lose:

Music as desolate, as beautiful
as your loved places.
mountainy marshes and glistening mud-flats
by the stealthy sea.
NORMAN MacCAIG, from 'Curlew'

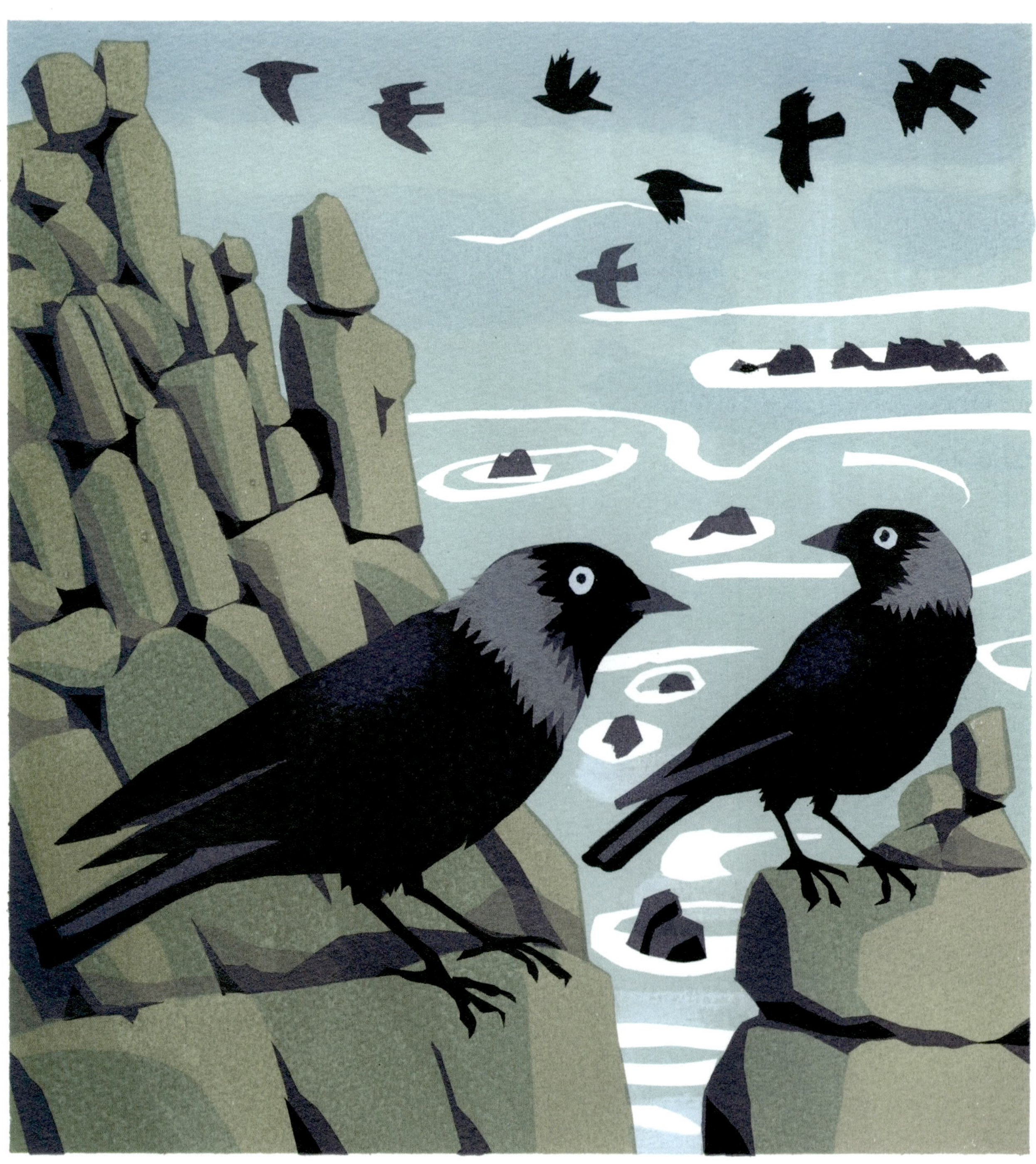

THE JACKDAW

Is there anything more joyful in nature than the heckle and clack of jackdaws, breaking across the sky with the sound of 15 struck snooker balls? 'Jack' is a precise imitation of their calls, as is 'daw', but jackdaw (*Corvus monedula*) derives from Latin *monedula*, the same stem as *moneta*/money, from its perceived love of bright things. Hence its appearance in the pangram, a sentence containing all the letters of the alphabet: 'jackdaws love my big sphinx of quartz'. Suitably its eyes are as brilliant as pearl buttons. Its thieving reputation was compounded by the most famous of the nineteenth-century Rev Richard Harris Barham's bestselling *Ingoldsby Legends*, the 'Jackdaw of Rheims'; the 'jack' in jackdaw thus open to roguish interpretation. In Barham's burlesque poem the cardinal briefly lays aside his ring to wash his hands after a feast. A jackdaw, which has been hopping about

Here and there
Like a dog in a fair

steals the ring. The cardinal curses everyone and everything under the sun. The jackdaw sickens from the curse and is duly exposed as the thief. He discloses the ring, the cardinal lifts the curse, the jackdaw flourishes and is canonised when he dies.

And on newly-made Saints and Popes, as you know
It's the custom, at Rome, new names to bestow,
So they canonized him by the name of Jim Crow!

Jackdaws like to nest in tree holes and rock crevices, notoriously in chimneys and legendarily in churches. William Cowper envies their indifference to the human comedy below.

Thrice happy bird! I too have seen
Much of the vanities of men;
  And sick of having seen 'em,
Would cheerfully these limbs resign

For such a pair of wings as thine,
  And such a head between 'em.
From 'The Jack Daw'

They make good pets. Becky Parkin (*Country Life*, 6/6/18) remembered her 'Jacko', who made off with an elderly lady visitor's 'very expensive dental plate'.

Old beeches are a favourite nesting site, the rank nest disclosing black-spotted sky-blue eggs. Often they will pluck fur and hair from grazing animals for lining the nest, which in chimneys can be several feet deep as they build it up from the first lodged twig. Carry Akroyd reckons they quite like a fire in the grate below, the smoke ridding their feathers of parasites. They render the same service to animals, frequently deer, picking the ticks even from nostrils and eyes. By the sea, they enjoy hanging in the updraft from a cliff. Customers of the café at Cornwall's Lizard Point can sit while jackdaws float motionless and within touching distance as they hold their stations in the breeze.

Numbers dipped in the 1970s but by 2019 had recovered to 1.4 million breeding pairs, seen throughout the UK except in relatively treeless north-west Scotland. Their principal diet is insects, with the remainder divided between vegetable and animal matter. A jackdaw winging its way with a struggling starling chick is my recent unpleasant memory. Jackdaws often mingle with rooks, and urban absence is a good indication of pollution.

Gavin Maxwell, author of *Ring of Bright Water*, when training prospective wartime commandos in the Highlands, said he would eat whatever they shot, to impress on them how to live off the land. He expected a delectable grouse or mountain hare. He got a jackdaw.

Ian Morton ('I'm all right, Jack', *Country Life* 16/11/18) listed some jackdaw curiosities: that when Adolf Hitler appropriated art in the 1930s he was derided as 'the jackdaw of Linz'; 'kafka' means 'jackdaw' in Czech; and it is the emblem of Malmesbury, whose inhabitants are called 'jackdaws'.

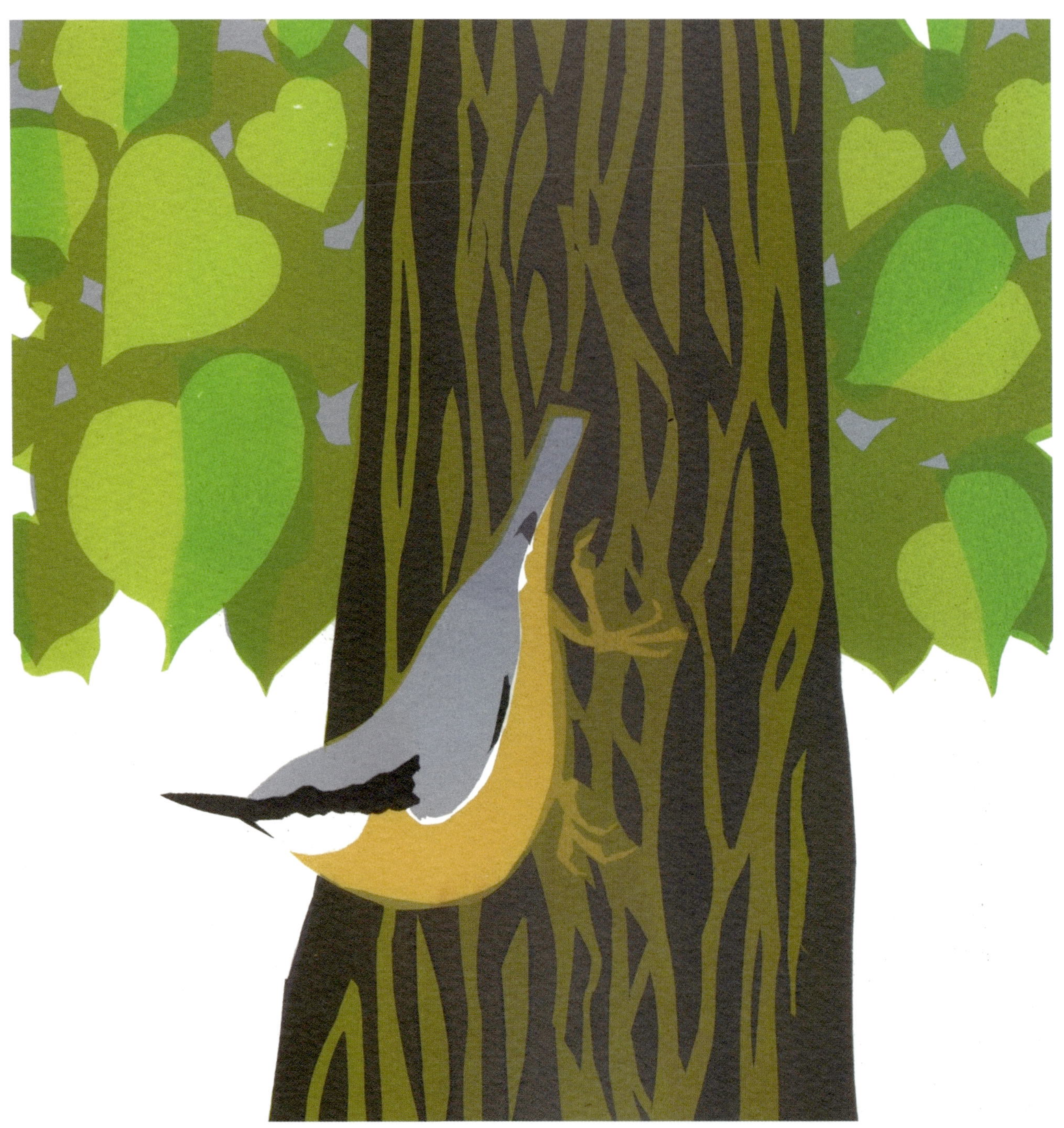

THE NUTHATCH

No woodland songster is more likely than the nuthatch (*Sitta europaea*) to provoke – by its full-blooded, far-carrying, lasso of a call – a 'Who the heck is that!', since the signature call could be a half wolf whistle. As W. H. Hudson wrote: 'Without being a songster in the strictest sense… his voice is so clear and far-reaching, and of so pleasant a quality, that it often gives more life and spirit to the woods and orchards and avenues he frequents than that of many true melodists.'

In summer showers a skreeking noise is heard
Deep in the woods of some uncommon bird
It makes a loud and long and loud continued noise
*   And often stops the speed of men and boys*
They think somebody mocks and goes along
And never thinks the nuthatch makes the song
Who always comes along the the summer guest
The birdnest hunters never found the nest
The schoolboy hears the noise from day to day
And stoops among the thorns to find a way
And starts the jay bird from the bushes green
He looks and sees a nest he's never seen
And takes the spotted eggs with many joys
And thinks he found the bird that made the noise.

JOHN CLARE, 'The Nuthatch'

Birds tend to cease singing at human approach; not the nuthatch. It can stand its ground and give full vent only feet away. The power of its unmodulated cry at close range is penetrating. The black eye-stripe, which divides its harmoniously tonal colouring, gives a bandit touch suitable to its boldness.

Persuading them to feed from the hand takes more patience. Even where birds trust strangers offering food, there will be a dozen visits from great tits to every one from a nuthatch. And whereas the great tit will linger long enough to swallow a couple of nibbles before swooping off, the nuthatch's flight is as straight as its call. It will arrive unannounced from a distance, pick the target, and zoom off to peck its prize at leisure. If trees are nearby they also come to bird tables and garden feeders.

Nuthatches are the only British birds with the squirrel-like ability to scurry down and around trunks and branches, as well as going up. Alister Warman who has a way of attracting avian surprises (see The Woodcock, page 29) looks forward to the day when he sees an eager nuthatch coming down a tree crash headfirst into a no less characteristically engaged treecreeper going up it.

The name, also more descriptively 'nut hacker', comes from their habit of wedging hazelnuts or acorns in cracks and then hacking them open. Gilbert White discovered that the 'jar-bird' referred to by Wiltshire rustics, which made 'a clatter with its bill against a dead bough or some old pales', was the nuthatch. He added that the 'noise may be heard a furlong or more away'. For a small, if chunky, bird, nut-hacking demands a super-strong beak.

They nest in holes in masonry and nest boxes as well as trees, using mud to reduce the entrance if required. It is wonderful to report breeding numbers are rocketing in England, Wales and now southern Scotland, having first been seen north of the border in 1989. Only Ireland remains un-nuthatched.

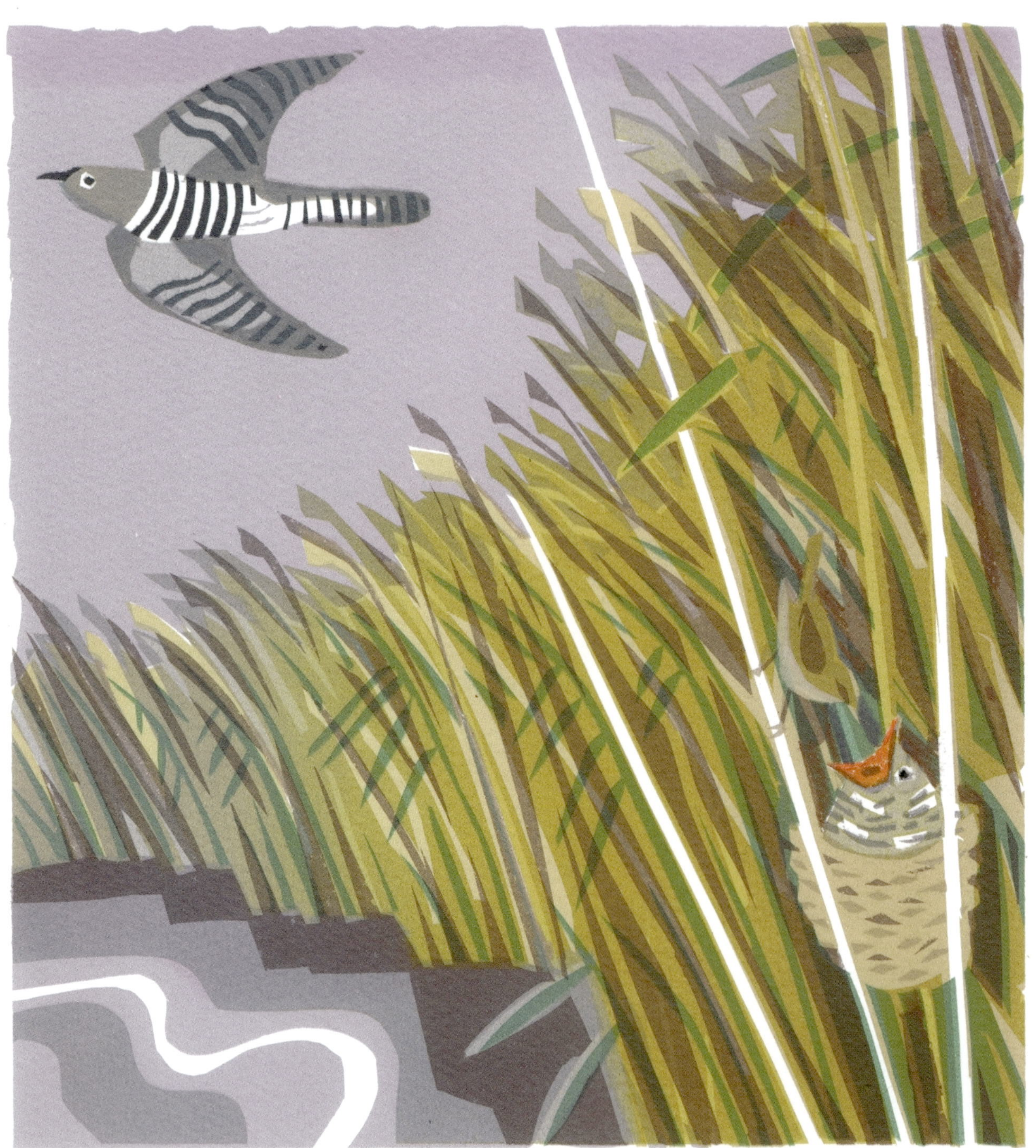

THE CUCKOO

'Hooray, hooray, the first of May, outdoor frolics begin today.' Hence the oldest scored song in English, the mid-thirteenth century:

Sumer is icumen in
Lhude sing cuccu

The male cuckoo (*Cuculus canorus*) 'cuckoos'; the female's call sounds like water jugging from a bottle.

P. G. Wodehouse (*Over Seventy*) saluted its literary importance by quoting Fiat Justitia, who had advised him that every great writer began with the cuckoo. The Master himself would surely have approved the index entry: 'Cuckoo, see Shaftesbury, Constance Lady' (John Wyndham, *Wyndham and Children First*). John Wyndham was prime minister Harold MacMillan's private secretary.

Tagging of East Anglian-based birds revealed one left in early June. It spent 47 per cent of the year in the Congo, 38 per cent on migration and 15 per cent in England. Welsh and Scottish birds travel via France, Germany, Libya and the Sahara; English ones via Spain, Morocco and West Africa, a route twice as dangerous.

A young listener asked Henry Douglas-Home: 'How is it that the cuckoo who has never seen its parents can grow up to say "cuckoo-cuckoo"? If the stork dropped a British baby in China, it wouldn't grow up speaking English, would it?' As he acknowledged: 'An unanswerable question.' Nothing has given the cuckoo a worse name than its 'cuckolding', laying its eggs in other birds' nests.

'Cuckoo, cuckoo!' O, word of fear
Unpleasing to the married ear!
WILLIAM SHAKESPEARE, *Love's Labour's Lost*

Edgar Chance, director of a Birmingham glass company and passionate egg collector in the days of former abundance, directed the pioneering wildlife film, *The*

Cuckoo's Secret (1921), which revealed for the first time a cuckoo laying its egg. Research since has proved this a strange and artful process, as Nick Davies's book, *Cuckoo: Cheating by Nature*, explains. The cuckoo's main British hosts are the dunnock (hedges, scrub, woodland), reed warbler (reed-beds) and meadow pipit (heath/moorland). The cuckoo appears to have been imprinted with the species that reared it. So a bird that was fostered by a reed warbler will lay an egg imitating a reed warbler's in a reed warbler's nest. The same with the meadow pipit. The dunnock contrarily does not require imitation, whereas the other hosts would eject a suspicious egg or abandon the nest. The cuckoo can wait for two hours or more, usually in late afternoon, for its opportunity, then acts swiftly, often encouraging the host to lay by previously emptying the targeted nest of its eggs. If there is an egg, it picks it up, lays its own, and leaves within 10 seconds. It may lay an egg in a dozen nests of the particular species. The monstrous chick, its back designed for the job, later completes the occupation by heaving out any remains of the victim's clutch, unborn or born; a gruesome operation first described by Aristotle.

British cuckoo numbers have nose-dived. Davies's favourite location, Wicken Fen near Cambridge, the UK's oldest nature reserve, was host to 14 females in 1985 and in 2012 failed to fledge a single one. I last listened to the call on Harris, a sound inseparable from childhood but previously unheard by my sons, and melancholy in that mountain fastness.

Strange, voice, unsatisfied,
*   whose echoes over all*
the empty countryside
*   repeat their urgent call*
…Whom do you thus beseech,
*   voice, calling endlessly*
down corridors of beech
*   and never a reply.*
JOHN H. F. McEWEN, from 'The Cuckoo

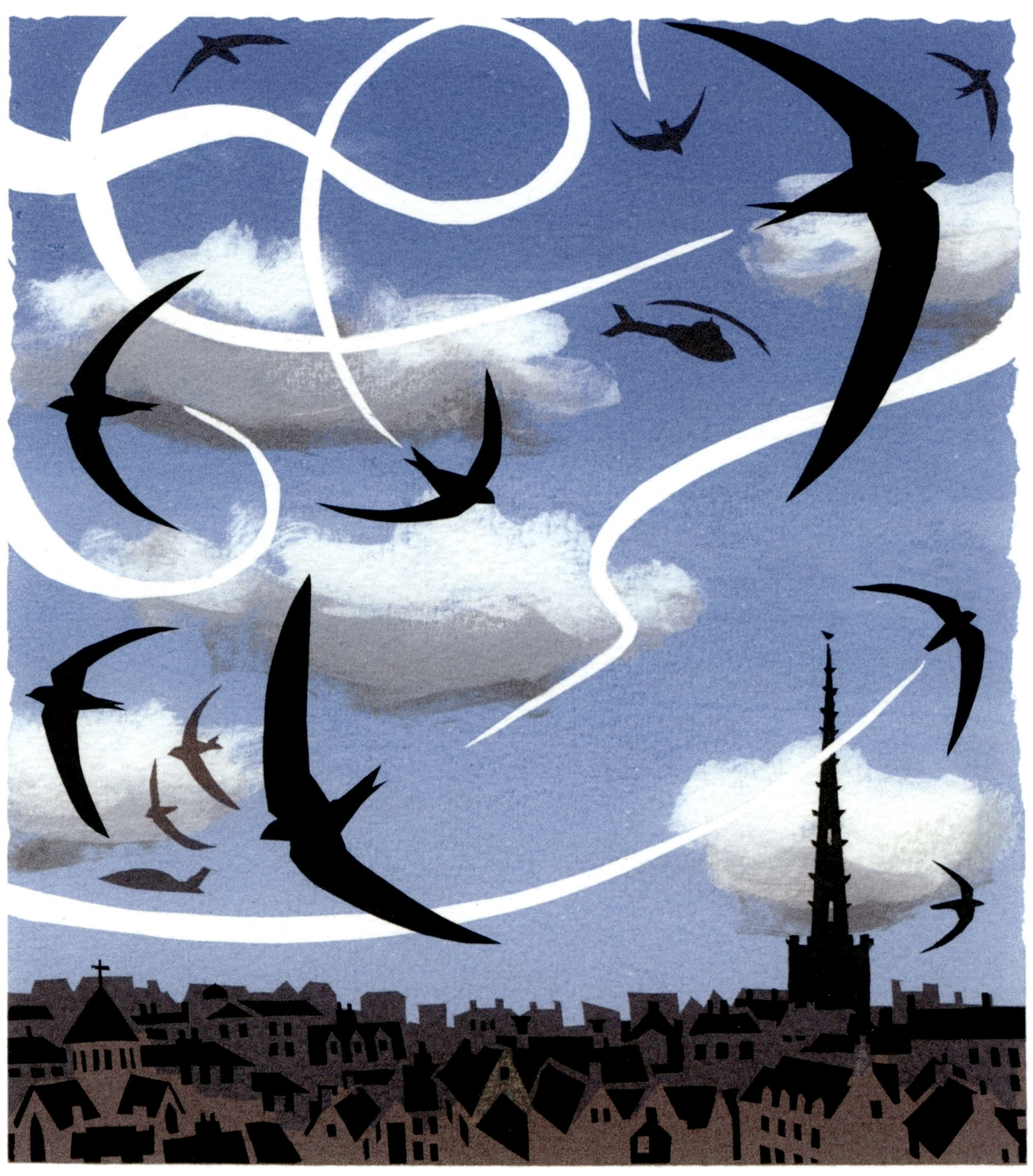

THE SWIFT

We measure the year by the astronomical calendar. The common swift (*Apus apus*) knows better: summer begins with May, not on 21st June, and ends with July. No bird is more eagerly awaited because the swift is the sound of summer, especially for town dwellers.

While over them shrill shrieked in his fierce glee
The swift with wings and tail as sharp and narrow
As if the bow had flown off with the arrow.
EDWARD THOMAS, from 'Haymaking'

Young birds intensify the departure, congregating for a week or two in teeming groups before they leave for sub-Saharan Africa around 1 August, leaving a silence as melancholy as their arrival is uplifting.

Henry Douglas-Home, pioneer of swift nest boxes, was first captivated as a schoolboy in the Thames Valley at Eton: 'When they stopped playing, just on dark, they all circled up screaming, up and up into the sky, fainter and fainter till they disappeared from sight and could no longer be heard. From that moment I decided swifts were the most fascinating and mysterious of all birds.' That swifts sleep at night on the wing was first confirmed by a French First World War pilot over the Western Front. They can cruise, far beyond danger, 2½ miles (4km) above the Earth.

The swift is biologically unique, with its own taxonomic family and order. Until they breed at four years old, they may be totally airborne. The availability of aerial food is indicated by the scientific calculation that in a typical British summer month 3 billion insects are airborne, many, including spiders, carried to heights of four miles (6.4km). A swift can live for over 20 years, thus totalling more than 3½ million flown miles. Its record speed is 106 mph; the 4-inch (10-cm) frame propelled by a 15-inch (38-cm) wingspan. The frail feet can grip but offer meagre support; hence Aristotle named it 'footless': a = without, *pous* = foot. In Poland's Białowieza Primeval Forest they nest, as they originally did,

in tree holes. The world's longest swift research project is at Oxford's Natural History Museum. It has revealed that in famines chicks can survive by lying torpid for up to 48 hours. At such times British-based swifts can roam as far as the Baltic, storing their catch as a congealed ball in their throats. During the wet summer of 2013, starved birds fell dead from the sky in Spain and France. In the heat of 2018, they were seen uncharacteristically dipping for water like swallows.

Swifts are in severe decline, not helped by laying only three eggs and having a single brood. London numbers have fallen 40 per cent since 2000. They like old churches (Carry Akroyd shows St Peter's, Oundle) and old buildings, which offer nooks and crannies for nesting. In 2013 Belfast launched its 'Swift City Conservation Project', making it Britain's first official 'swift city'. There are many such initiatives in the UK and Europe, not least in the roof of London's second-tallest building, 1 Canada Square, Canary Wharf. Legally enforced swift-box provision – some designed to merge with brickwork – is required.

Summer without swifts is unthinkable.

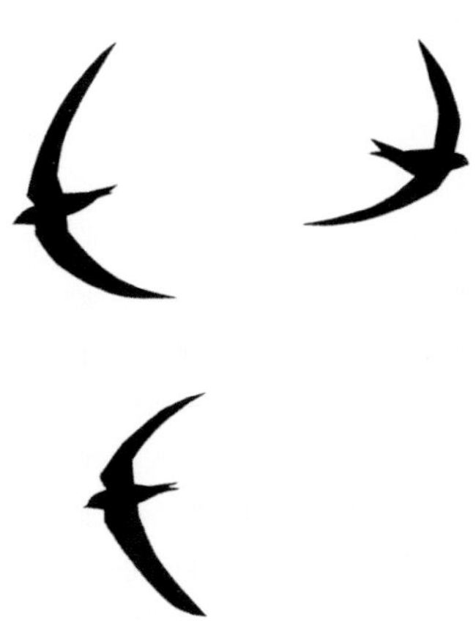

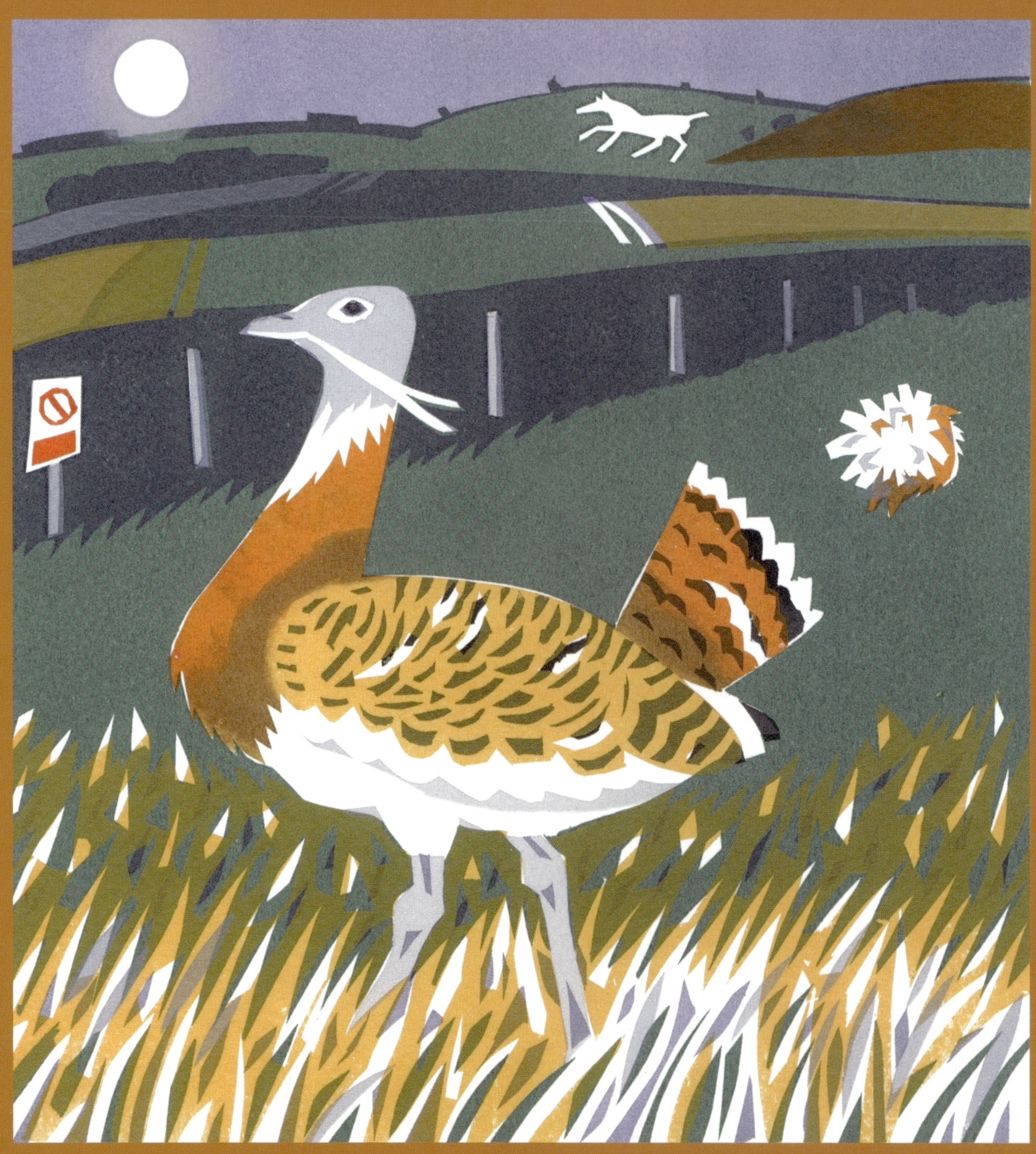

THE GREAT BUSTARD

'Of all the regular British breeding birds in the last two hundred years, arguably only one, the great bustard, no longer breeds in Britain', wrote David Waters, Executive officer of the Great Bustard Group, and Estlin Waters (*The Great Bustard*, Great Bustard Group) in 2006. One hundred years before that W. H. Hudson wrote: 'The order Alectorides, which follows, includes two noble forms once common, but now extinct in this country. One is the crane (*Grus communis*)…The other, finest of British birds, is the great bustard (*Otis tarda*), which lived in all suitable localities in England, from the southern counties to Yorkshire, and was wantonly extirpated during the first half of the present century.' Just how widespread the bird was before the industrial age is confirmed by its presence in Wiltshire's coat of arms and twice in Cambridgeshire's, which, as Carry Akroyd says, 'obviously needs the attention of the Great Bustard Group'. Hudson would surely be amazed to learn that in 2018 both birds have a nascent wild population in England.

One morning after breakfast in the third week of April 2018, Carry and I visited the GBG's main reserve on Salisbury Plain at the invitation of Simon Courtauld, suitably a former editor of *The Field*. We hoped to see the male's legendary courting display. In the spacious hide we were warned that two Mancunian enthusiasts had recently made the pilgrimage twice to no avail. The fenced-off reserve on the hillside opposite, which offered substantial cover, contained a male and, having briefly watched its well-camouflaged progress, the transformation took place.

The slender, earth-fused, bird stopped – and slowly became a dazzling white-feathered globe, which beamed its magnificence for several minutes (usually eight maximum) before disappearing as suddenly as a switched light. The bird achieves this 'balloon' display by gulping air to inflate its feather-ruffled neck and forming an enclosure with reversed spread wings and fanned tail, the whole crested by now erect white whiskers. The display, which is non-territorial, is first seen in spring but can continue into June. It usually takes place in the early morning, but also occurs on moonlit nights – the ultimate wonder.

Great bustard males weigh up to three stone (19kg), five times heavier than the female, and are a foot taller at 3½ feet (107cm) – the greatest gender difference in size of any bird. The male's wingspan can be nine feet (2.7m). The bird's average life expectancy is 10, the record 28. Males tend to die younger than females, which do not breed until they are three or four, while males mate at five or six. Great bustards are omnivorous, mostly vegetarian although their diet can include small reptiles and mammals. They often migrate in flocks, reaching a speed of 50 mph. A bird came gliding in during our visit, and another took off effortlessly as we left, beating its way steadily into the distance, long legs tucked out of sight. The reserve is not a zoo but is successfully fenced against foxes and badgers. Recently installed Good Nature traps are hoped to deal with stoats.

The last great bustard nested in England in 1832. The bird's reintroduction to its ancient stronghold of Salisbury Plain was the unrealised dream of England's foremost wildlife-art gallery owner, Aylmer Tryon. His 1970s Great Bustard Trust was succeeded in 1998 by the GBW, a volunteer-dependent charity licensed to research by the government ministry DEFRA. The project was set up on the edge of the military training area. Chicks and eggs were imported from Russia and Spain. The first annual release took place in 2004.

There is a total population of 35,000 great bustards. 2018 was the GBT's best year: eight nesting females were identified, successfully producing four chicks. Four eggs were taken from two nests on the reserve – young females tend to cope better with one chick – and were hatched and reared. It was the first time eggs were not imported. The rearing team have to disguise themselves as bustards. Garbed like astronauts, they deliver food via pincers in the form of imitation bustard heads. The UK population of 70 birds is now self-sustaining. A drove of 29 was seen on Salisbury Plain (11/3/19).

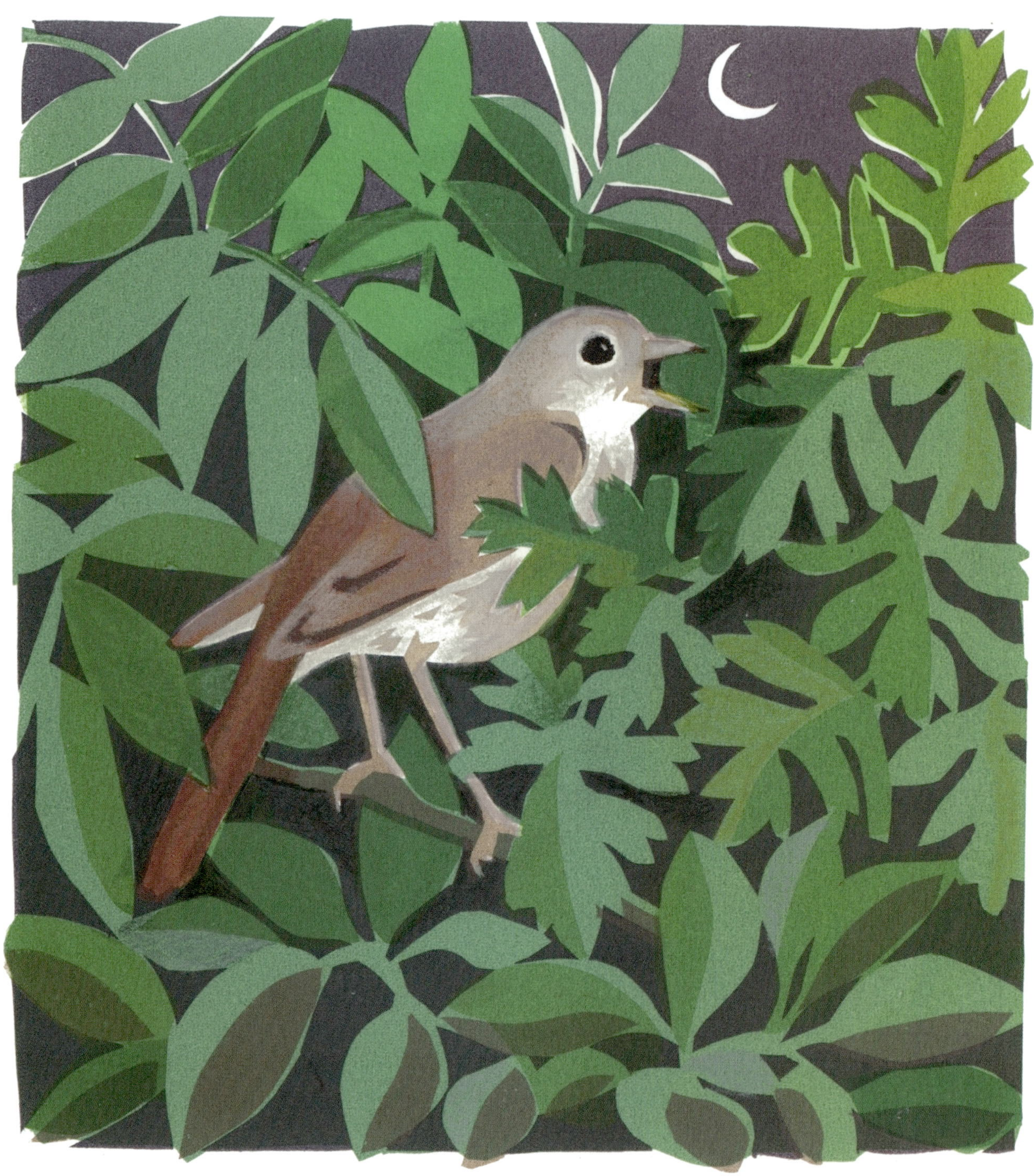

THE NIGHTINGALE

ALFRED, LORD TENNYSON, from 'Aylmer's Field'

The song of the nightingale (*Lusciana megarhyncos*) 'reverberates like a chord through European and Asian poetry', wrote the poet Edward Hirsch. Izaak Walton wrote that its song 'might make mankind to think miracles are not ceased' (*The Compleat Angler*). On the Western Front in the First World War, nightingales went on singing despite bombardments (John Lewis-Stempel, *Where Poppies Blow*). The BBC's first live outside broadcast (1924) featured cellist Beatrice Hamilton – also first to record Elgar's Cello Concerto – accompanying a nightingale in her Surrey garden. A million listeners tuned in. In the present century, folksinger Sam Lee has replicated this event, singing to the accompaniment of nightingales and musicians for an audience on the South Downs as part of the annual Brighton Festival. John Clare imitated the legendary song in verse:

'Wew–wew wew–wew chur–chur chur–chur
Woo–it woo–it' – could this be her?
'Tee–rew tee–rew tee–rew tee–rew
Chew–rit chew–rit' – and ever new –
'Will–will will–will grig–grig grig–grig.'

From 'The Progress of Rhyme'

Nightingales (Old English, 'nightsingers') have 200 distinct songs, most singularly sung at night (23½ hours non-stop the record), in late April and May after its summer arrival from equatorial Africa. In 'To the Nightingale', John Clare marvelled such a plain 'clod-brown' bird could sing 'so fine a song'.

In old poetry, it is often called Philomel, from the myth of Philomela. Philomela was reborn a nightingale after her tongue was cut out on the orders of her rapist brother-in-law Tereus,

King of Thrace. Hence the origin of the sorrowful nightingale: *Luscinia*, from *luctus*, 'lamentation', and *cano*, 'to sing'. Clare would have agreed with the poet Sappho 2,500 years before, who wrote 'spring's messenger, the sweet-voiced nightingale'. So would the rejected lover of 'A la claire fontaine':

Chante, rossignol, chante,
toi qui as le coeur gai

The nightingale is not the only nocturnal songster. Blackcaps are called 'northern nightingales' and sedge warblers 'fishermen's nightingales'. Artificial light can dupe birds into thinking night is day. At midnight on 4 January 2019, a robin greeted me home in Camden Town. London nightingales were once so numerous Edward the Confessor complained they disturbed his prayers. But in Second World War London, Eric Maschwitz was right when he wrote 'A Nightingale Sang in Berkeley Square'. Only the love-struck could imagine such a miracle. The bird is claimed to have another miraculous power: Tom Cruise, Victoria Beckham *et al.* apparently swear by anti-ageing facials of Japanese nightingale droppings.

The diminishing 3,000 British breeding pairs are confined to the south-east, principally Kent and Essex. The most famous nightingale remains the bird John Keats heard one night in Hampstead:

Now more than ever seems it rich to die,
To cease upon the midnight with no pain,
While thou art pouring forth thy soul abroad
In such an ecstasy!

From 'Ode to a Nightingale'

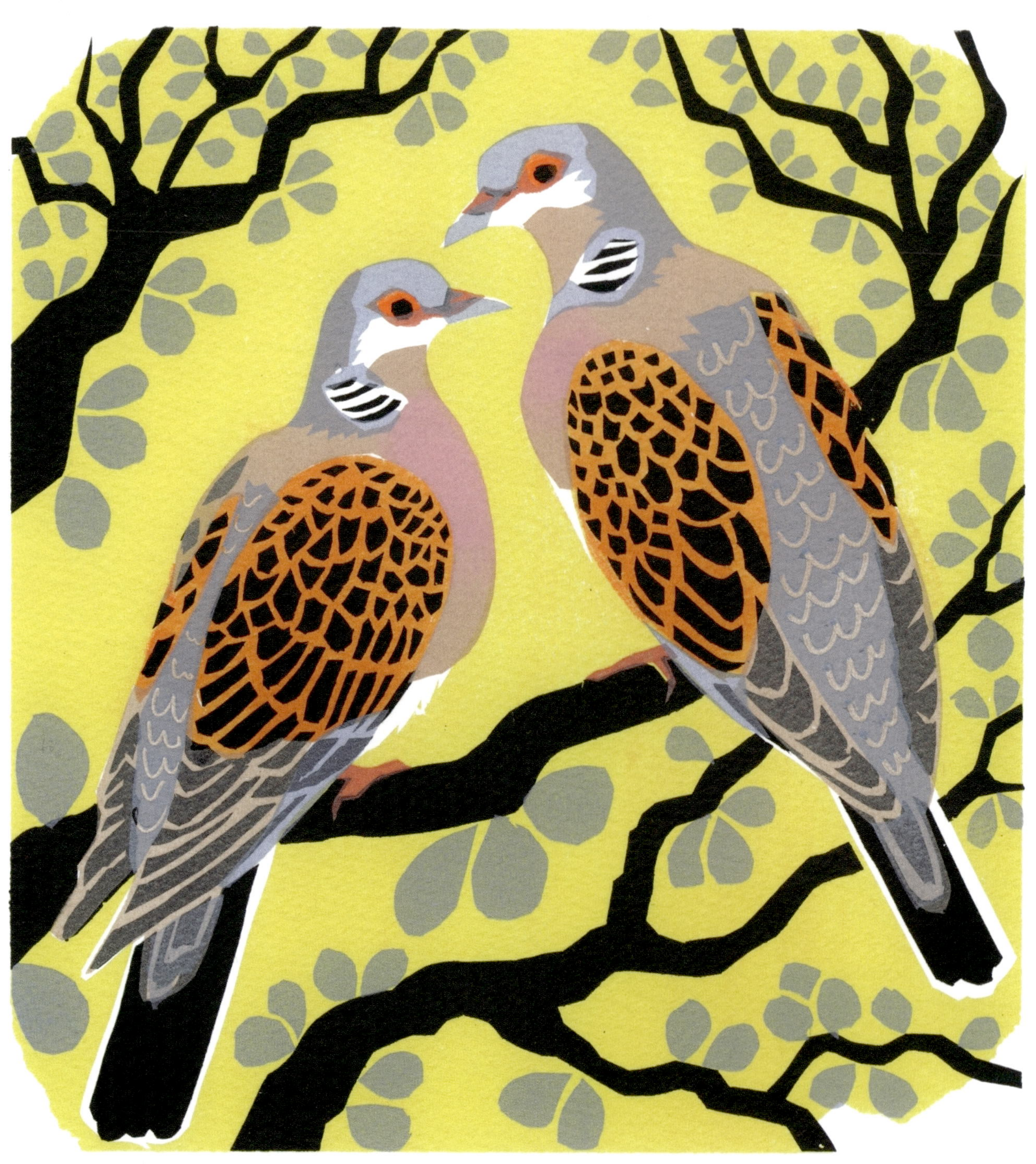

THE TURTLE DOVE

The turtle dove (*Streptopelia turtur*) derives its name from its purring 'turr-turr' that chimes so perfectly with a cloudless summer's day.

For lo, the winter is past, the rain is over and gone;
The flowers appear on the earth; the time of the
singing of birds is come, and the voice of the turtle
is heard in our land.

The Song of Solomon 2:11–12, *King James Bible*

This shows that before the world became imperial Britain's oyster, and turtle soup a consequent delicacy, 'turtle' (from onomatopoeic Latin *turtur*, French *tourterelle*) in common speech referred to our only migrant, and smallest, pigeon. By the eighteenth century, with trade routes well established, things had changed. The turtle (dove) still took precedence in Dr Johnson's *Dictionary of the English Language*, but it had a rival: '*tu'rtle*: It is used among sailors and gluttons for a tortoise.'

This prettiest of UK doves has symbolised marital fidelity, lending weight to 'lovey-dovey'. Nothing better symbolises this than its ecstatic courtship flight, an exotic version of the woodpigeon's, with rapidly beating wings and a gliding, often circular, descent, wings and tail spread wide. They mate for life.

The most famous poem in its praise is Shakespeare's 'The Phoenix and the Turtle', but poetic recognition of its constancy already appears anonymously in England in the thirteenth century:

By her mate she sits a-night
A-day they goeth and flyeth;
Whoso says that they are sundered,
I say that he lieth.

John Gay in *The Beggar's Opera* repeated the theme of their fidelity four centuries later:

The turtle thus with plaintive crying
Laments her dove.
Down she drops quite spent with sighing,
Paired in death, as paired in love.

One cause for lament is the bird's popularity as a dish. It was already trapped by the ancient Egyptians, and the slaughter continues today when the bird travels to and from its Eurasian summer breeding grounds to winter across southern Central Africa – where, as recently as the 1970s, its congregations numbered up to a million. Britain's 'Birds of Conservation Concern' assessments over the 20 years prior to 2019, found the turtle dove to be the most abundant species to be assessed as critically endangered – due to a 91 per cent decline over the last three generations (16 years).

Despite the continuing migrational slaughter by mist nets and guns in North Africa as well as some Mediterranean disregarders of conservation laws, habitat reduction and a protozoan parasite probably pose the gravest threat. Water on migration is vital; suitable seeds – all doves have more delicate metabolisms than the woodpigeon – have to be near the nest for young to survive.

The rewilding of the previously farmed Knepp Castle Estate in West Sussex has attracted probably the only new colony of turtle doves in Britain and their greatest current density – 16 singing males in 2018; but Isabella Tree, author of *Wilding*, who owns the Knepp Wildland Project with her husband, Charlie Burrell, says isolated schemes will not work as the birds have nowhere to spread; the UK being the most densely peopled of major European countries. Hearing a Knepp bird, she concluded: 'The gentle mournfulness of its call seemed' its 'swansong'. By happy contrast, at Weyerburg in rural Austria on 16 June 2018, I was awoken before dawn by the singing of a nightingale, followed with the coming of the light by a cuckoo and finally a turtle dove.

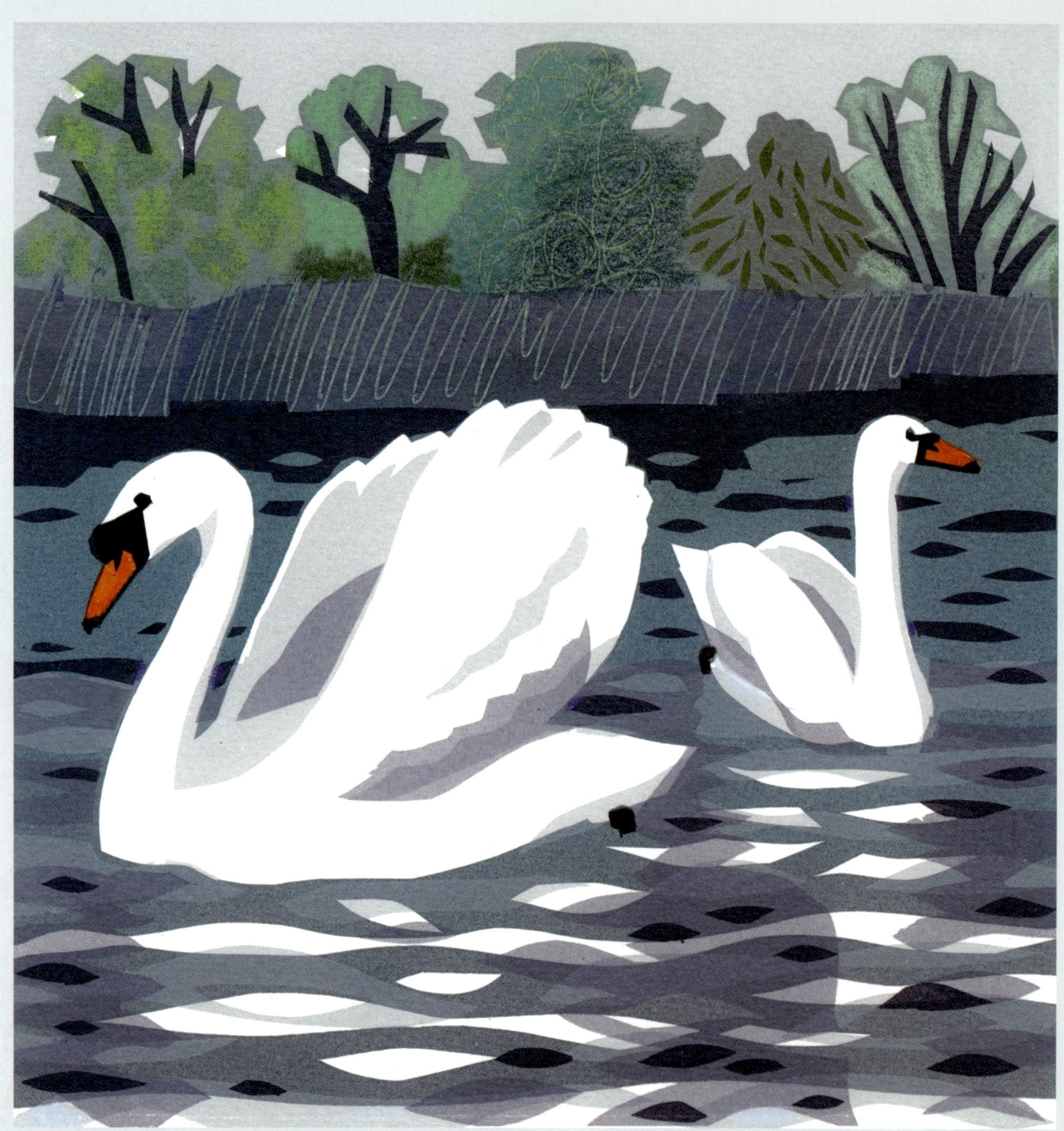

THE MUTE SWAN

In his references to 60 different birds, Shakespeare cites the mute swan (*Cygnus olor*) more than any other. When Ben Jonson memorialised him it was the mute he chose as the Bard's symbol:

Sweet Swan of Avon! What a sight it were
To see thee in our waters yet appear,
And make those flights upon the bankes of Thames
That so did take Eliza and our James!

From 'To the Memory of my Beloved the Author, Mr William Shakespeare'

The mute is the summer swan of ponds and lakes, its spinnaker wings surely make it the balletic inspiration of Tchaikovsky's *Swan Lake*. Cob and pen have black knobs at the root of orange bills and hiss and snort when threatened. They are 'mute' indeed compared with the clarion-calling winter migrants, whooper and Bewick's swans. Unlike these counterparts, the mute's wing beats throb noisily once it has thrashed its weight (up to 40 lbs/18kg) airborne with the aid of an 8ft (2.4m) wingspan.

Abbotsbury in Dorset is the world's only managed colony of mutes, its foundation dating from the eleventh century. In July, during the six-week annual moult, there is a bi-annual round-up to ring and examine its 600-odd birds. July's third week sees the similarly historic operation of Swan Upping on the Thames – from Sudbury to mid-Oxfordshire. Today the birds' Thames owners are the sovereign and (since the fifteenth century) the Vintners' and Dyers' livery companies. Unmarked swans were traditionally Crown property, mutes the ultimate roasting bird until the nineteenth century, with cygnets 'upped' to be fattened for the table.

The sovereign is Seigneur of the Swans. The swan uppers, in bright livery with a mute's quill in their caps, man a variegated flotilla under the direction of the Royal Swan Marker. The recorder is Royal Swan Warden. Clive Aslet wrote of the ritual: 'I recommend it to anyone who's in danger of losing faith in the modern world' (*Country Life*,

7/8/14). On 20 July 2009 the Queen attended, the first sovereign to do so in modern times. The tradition that unmarked swans are Crown property has long afforded them protection.

Since the 1987 ban on lead fishing weights that, when swallowed, caused birds to die from poisoning, the mute population has doubled and the Thames breeding pairs have trebled. In winter, Britain's 35,000 population flocks and is doubled by winter migrants.

Hans Andersen's autobiographical tale *The Ugly Duckling* features a mute – its ugliness the year it suffers as a mud-coloured cygnet. Adults can be brutally territorial. There was a Canada goose with deformed wings which lived in the bird sanctuary on Hampstead Heath and was subjected to furious bullying by a pair of semi-resident mutes. My wife's call of 'Goosey' would bring it from the farthest end of the pond to be fed, only for its tormentors to appear and chase it relentlessly until it had to escape drowning by scrambling on shore. Eventually, it had to be removed to a safer haven elsewhere. Ducks and swans have penises, so the myth of Zeus ravishing Leda in the guise of a mute is all too credible, as Yeats's erotic poem celebrates.

Mutes mate for life (40 the oldest age in captivity). A dawn-chorus outing with my wife after Goosey's removal, revealed our hated bullying couple, the cob floating asleep by its sleeping and incubating pen, an inspiring symbol of fidelity:

And wheresoever we went, like Juno's swans,
Still we went coupled and inseparable

WILLIAM SHAKESPEARE, *As You Like It*, Act 1, Scene 3

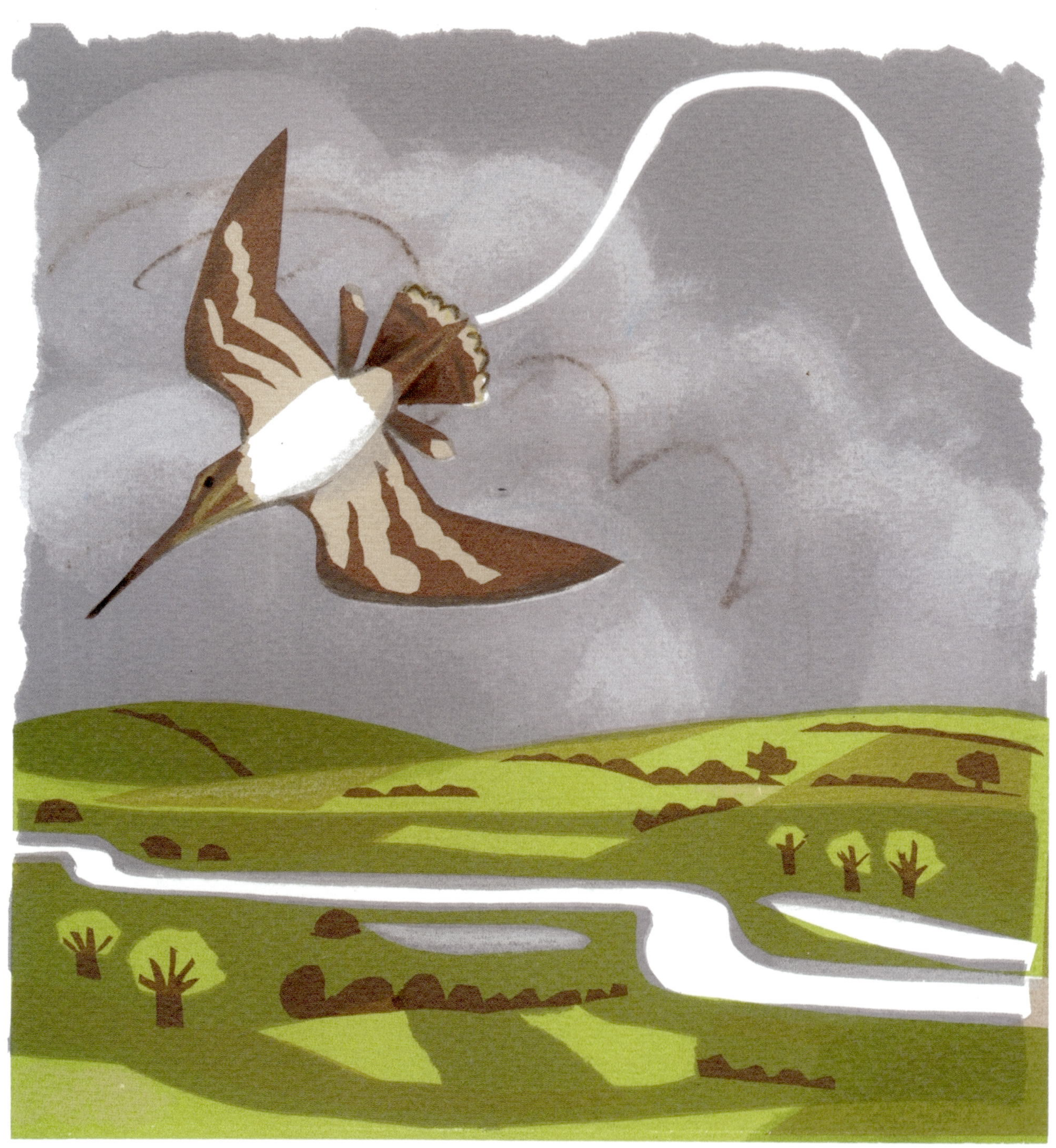

THE SNIPE

The woodcock goes on its 'roding' rounds in the spring twilight. Its cousin, the snipe (*Gallinago gallinago*), has an equivalent but more exuberant daylight sound – not a call, like the woodcock's grunted ruminations. The sound, often blending with the bleat of lambs in hill country, can puzzle a listener. But the sky will soon reveal the speck of a bird, continuously rising to a height then plummeting earthwards, the wind from its dive making the outer pair of tail-feathers, spread at rigid right angles, vibrate noisily; the same principle used by terrorising German Stuka bombers in the Second World War. It is usually referred to as 'drumming', which describes the physical action but not the effect.

In westerly and northern Gaelic-speaking country, prime snipe-bog country, they call the bird 'the goat of the air'; although only the male creates the 'bleat' as it dive-bombs the breeding ground. The cause was long disputed until, in a 1912 lecture to the British Ornithologists' Club, Philip Manson-Bahr demonstrated its origin: he stuck two snipe outer-tail feathers in opposite sides of a cork, then whirled it around on a length of string (*Birds Britannica*). All snipe, when put up, also make a rasping, single-syllable alarm call. They are tight sitters, so it is the call which usually signals their presence. They can accelerate to a searing 60 mph, swerving this way and that to evade a hawk or a hunter's shot.

Always a little startling, this 'jump' of a snipe; it seems to spring out of the ground, to be quite invisible until it is in the air. Even when you see a snipe alight (with the same suddenness), as soon as it touches the earth it seems to 'snuff out', so cunningly does it harmonise with its surroundings. The 'jazz' pattern on its back is responsible.

'BB', *BB's Birds*

The woodcock's plumage blends with leaf litter, the snipe's zig-zag 'jazz' streaks with bent dry-grass stems.

Lover of swamps
And quagmire overgrown
With hassock–tufts of sedge, where fear encamps
Around thy home alone

JOHN CLARE, from 'To the Snipe'

Its camouflaged concealment gave rise to the hidden 'sniper' sharpshooter, hence 'sniping'. 'Gutter-snipe' is a Northumbrian name, otherwise applied. 'Snipe' was a nickname for lawyers, their bills notoriously long. The snipe's, at 3 inches (7.6cm), is almost a third of its body length, its soft tip an engineering marvel of microscopic sensitised pads, which allows it to locate its sub-mud prey.

In winter the UK's 80,000 resident breeding pairs are, like the woodcock's, raised to a million by continental migrants. Winter is the time to see them, when they gather in flocks or 'wisps' and feed on the foreshore. They are a prized delicacy and quarry; an easier target when driven to fly over stationary guns than is the famously difficult swerving bird for the walking rough shooter.

Churchill liked a brace cold for breakfast, with a pint of port. Sir Alec Douglas-Home liked one for breakfast – cold, with a poached egg, and tea. 'Always tea, nothing stronger. Most guests were revolted by the snipe – with the beak poking through its chest!' recalled his eldest daughter, Caroline Douglas-Home.

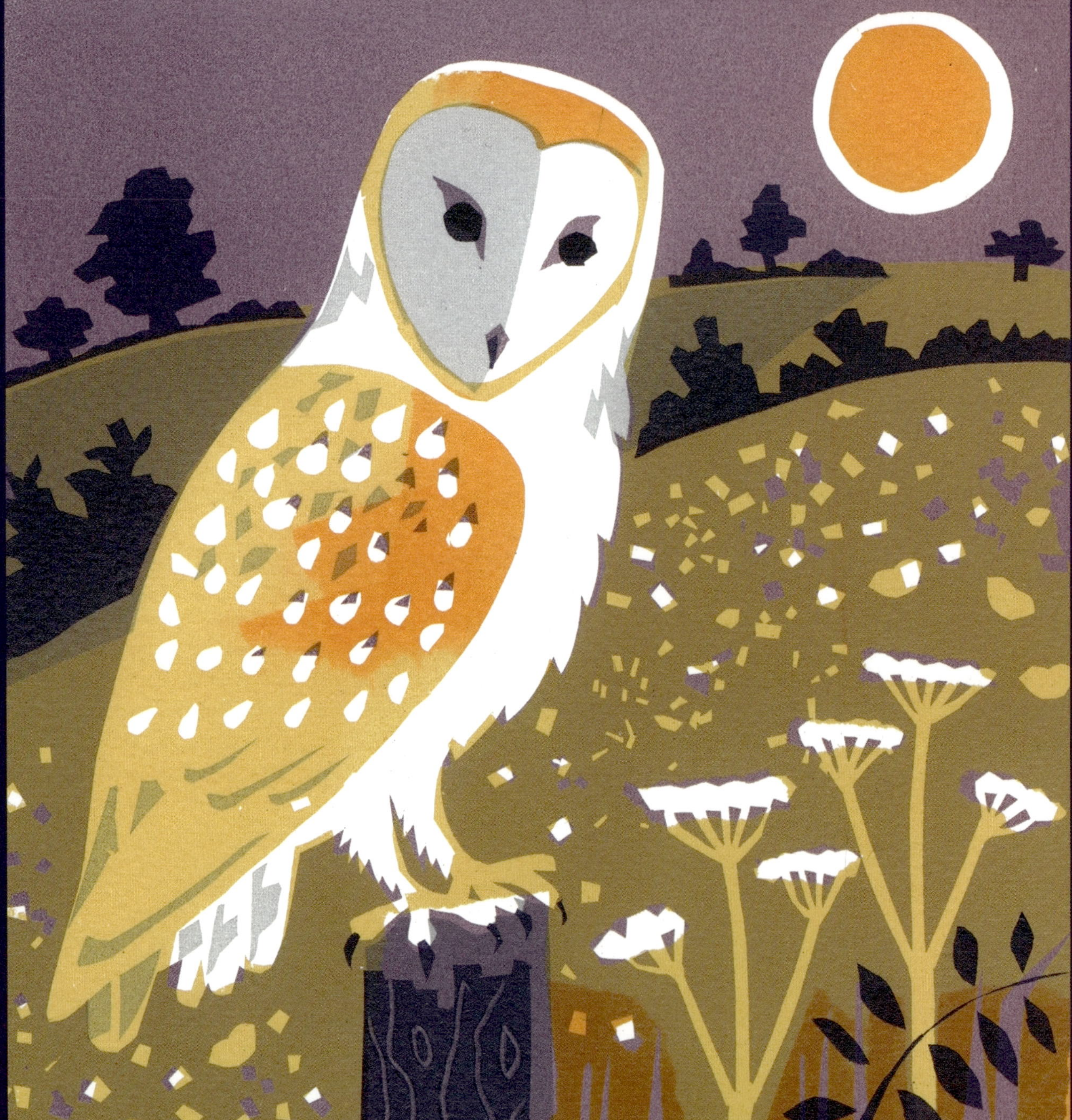

THE BARN OWL

Owls frontal-eyed faces act like satellite dishes, with sound-sensor hairs around beak and ears, aided by noiseless flight. Britain's second-favourite bird, the barn or white owl (*Tyto alba* (white)) is the most silent avian flyer. There is no more dreamlike midsummer sight than its ghostly hunting at dusk.

Owls' flight-feathers have noise-reducing serrations and down, especially the barn's; its wing weighs 0.007oz (210mg), a tawny's 0.01oz (400mg). Lack of waterproofing makes rain its enemy. It has the slowest, least vibrant, wingbeat yet great agility, including vertical take-off. Geoffrey Lilley (1919–2015), 'father of aeroacoustics', studied owl wings to reduce jet-engine noise and aircraft turbulence.

And be it a bird softlier fledge
Than white owl or brown nightjar,
Be softer the down on the wing's edge
Than combing crests of a snowdrift are
Which the smooth wind holloweth,
Of its shadowing I will be more aware
Than a mirror is of a swoon'd man's breath…
LASCELLES ABERCROMBIE, from 'An Escape'

In 1965 I adopted the weakling of a Berwickshire brood. 'It is composed of pale fluff, so soft that it is impalpable to the touch, and has an immensely sad, ancient appearance,' my mother noted on 19 June. She named it Melchior after one of the three wise men; gendering barn owls is for scientists. My neighbouring brother and sister-in-law had a tawny owlet, a fluff ball that pocketed all the compliments.

I fed my reptilian pet furred, raw rabbit until it lay lifeless and distended. The problem was a first for the vet, but his injection of liquid paraffin worked. Back home Melchior somehow disgorged a pellet the size of itself. From then on it was nibbles. By 26 June it was pattering about and hissing. Soon tables were turned with the tawny owlet. 13 July: 'Melchior is now very beautiful and friendly with his dark eyes in his moonlight face.' It solaced my widowed mother, purring in her lap when stroked. Un-house-trainable, it did give strenuous notice of excretions, and pellets are dry, but trial flights required an outdoor hutch. This brought a glimmering flock of barn owls to the garden. 14 July: 'I can hear their strange snoring and escaping-steam-like noises around the house at night.' Newspapers said there was a disastrous national decline. A fair proportion of the supposed Scottish population was in this one garden. I have been sceptical of bird counts ever since. Today the Barn Owl Trust offers vital assistance.

3 August: 'Melchior has found his wings and took off for the roof. I hope this is not the end of our delightful friendship.' Visits dwindled and finally ceased in October. Coincidentally a dead barn owl was found on the main road a mile away, a common occurrence when roadkill offers young owls easy pickings.

Betty Stokes and Alise di Robilant, friends from Italy, had spread the word. Melchior is memorialised in Giulia Niccolai's novel *Il Grand Angelo*: *'Quando vede la pallototola la nanny e contenta e dice*: "He's done his ball of pelt! He's done his ball of pelt!"'

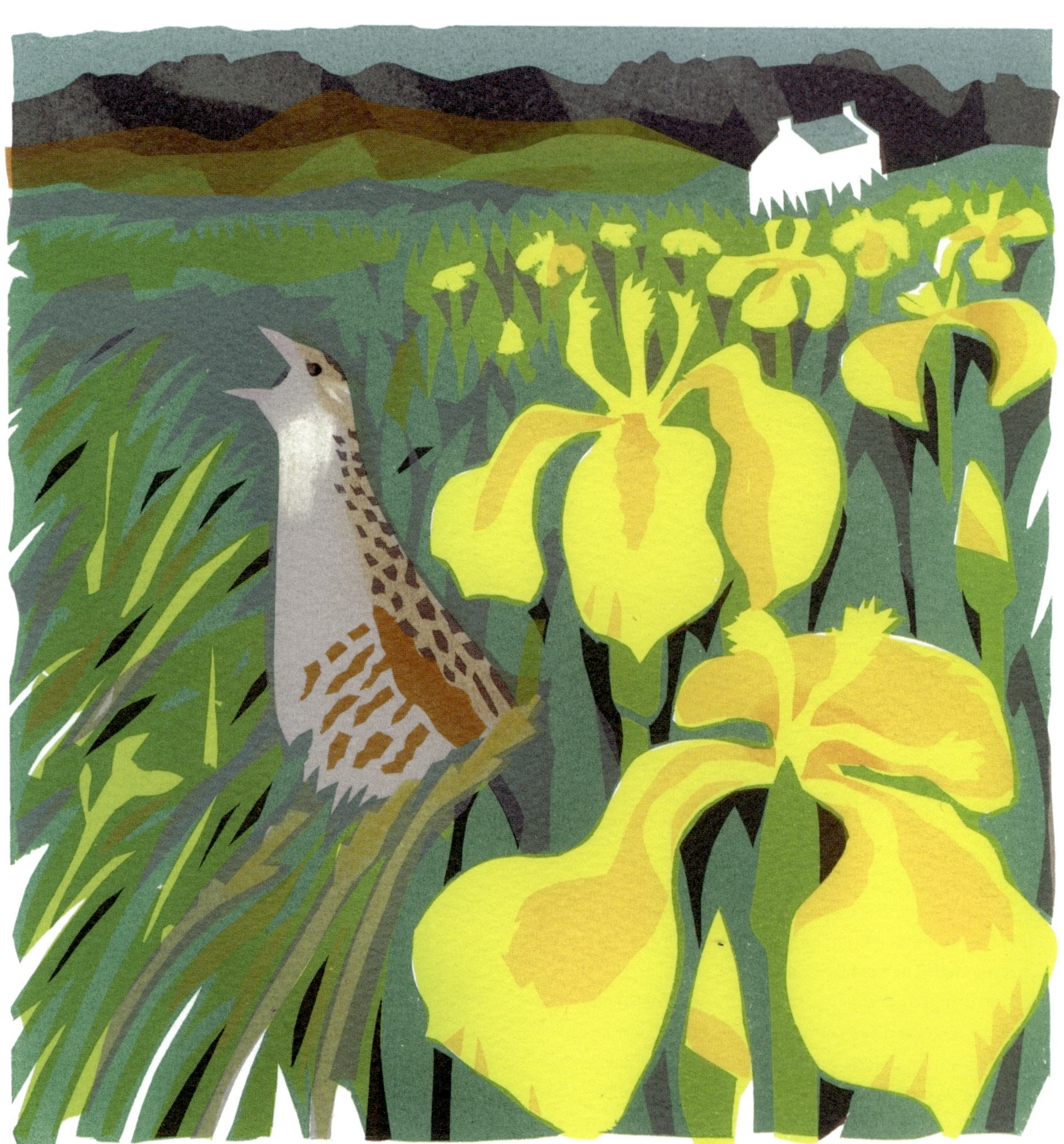

THE CORNCRAKE

The corncrake's Latin name, *Crex crex*, echoes the rasping and insistent call of the males by day and especially night, most regularly in May and June. 'The call,' my brother Jamie primed me in Ayrshire 45 years ago, 'is like this', and he dragged a finger along his comb. That June night, goaded by the call – a common reaction, I subsequently learnt – I rose from my bed and followed it to its source in the valley. The nearer it seemed, the more baffling its location, until I was standing in an echo chamber. I could have been surrounded by corncrakes. I returned to bed in the dawn confounded.

> *And still they hear the craiking sound*
> *And still they wonder why*
> *It surely cant be underground*
> *Nor is it in the sky*
> JOHN CLARE, from 'To the Landrail' (a traditional name for the corncrake)

It is by the voice we know them. Corncrakes skulk in cover and, even if seen, prefer to scurry than fly.

> *In this one of all fields I know the best*
> *All day and night, hoarse and melodious, sounded*
> *A creeping corncrake, coloured like the ground,*
> *Till the cats got him and gave the rough air rest.*
> NORMAN MacCAIG, from 'A Voice of Summer'

Their unwillingness to fly is deceptive. Corncrakes migrate from sub-Saharan Africa as far south as Madagascar, and have been found among the nocturnal kills of Bath's streetlight-aided peregrines; witness to their preferring night flight on migration. The Bath peregrines suggest there may be more corncrakes than experts think, but there is no denying a headlong decline. Lord Cockburn (1779–1854) judge ('except that he murdered, Burke was a gentlemanly fellow') and much else, heard them as commonplace in the heart of Edinburgh's New Town. Numbers fell across Europe as machines replaced scythes, and silage, hay. Since my corncrake chase, the population has declined by nearly three-quarters (UK: 1,000 breeding pairs) and in Ireland by 90 per cent. The only twenty-first-century record of a London sighting was at the Wetland Centre, Barnes, in 2010, the bird quickly despatched by a heron.

Today, the most reliable places to hear Britain's corncrakes are the Scottish islands and in County Donegal and West Connacht in Ireland. Coll is a particular stronghold, Tiree another, bought by the RSPB in 1991. Coll's grasslands are managed for corncrakes, with late-cut meadows and pasture left free from summer grazing. Protected areas, planted with cow parsley and nettles, provide early summer cover. Patches of yellow iris, such a feature of the Western Isles, are another favourite haunt. RSPB/Project Trust tours enable corncrake seekers to visit over two weekends in May and June. State-funded conservation schemes encourage farmers to be corncrake-friendly on other islands: grass is left to stand until August, fields are mown from the centre with cover left at field edges. 2015's fine summer on Coll resulted in a record 89 calling males. In England, since 2002, corncrakes have been reintroduced to the Nene Washes, the RSPB's reserve near Peterborough, almost to the place where John Clare heard them.

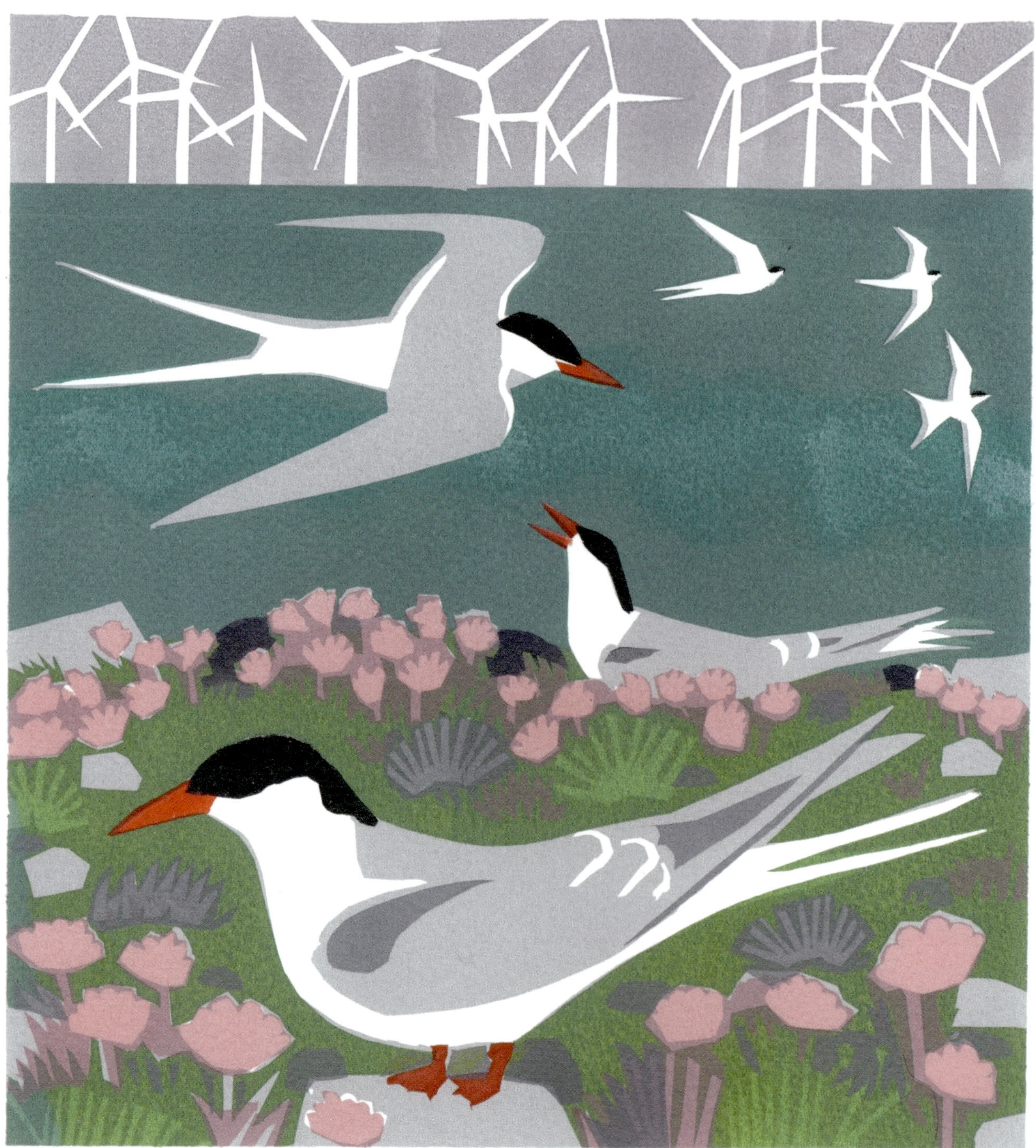

THE ARCTIC TERN

The majority of Arctic terns (*Sterna paradisae*) migrate from pole to pole, the longest journey of any living creature. In the summer of 2011 light loggers – sunlight-sensing geo-locators the weight of a paperclip – were retrieved that showed that five of the birds, fitted in Holland the previous year, had travelled an average 56,000 miles (90,000km) or the rough equivalent of a quarter of the way to the moon. This remains the avian record. Ring returns prove that an Arctic tern can live for 30 years. Not surprisingly it has an extraordinary physiology, weighing barely 4 ounces (113g) yet with a wingspan of 2½ feet (76cm); its body propelled by little more than a heart held by wing-muscle.

Its grace explains the paradisal name. The forked tail has earned it the description 'sea swallow'; the apparently effortless flight matching its white beauty and elegant trim. It wanders the Atlantic but can be seen inland, fishing lakes and reservoirs on its spring and autumn passage. Its ceaseless but leisurely circuits are instantly recognisable before it hovers and plunges to catch a fish. For protection it usually nests in colonies and, in common with other seabirds, a suddenly exposed shoal out to sea can provoke a feeding frenzy.

Linnaeus initially classified the Arctic tern *Sterna hirundo* (swallow tern), today the common tern's classification. This switch is in keeping with species so difficult to tell apart that unsure twitchers refer to 'comic terns'. The common tern has a black tip to its scarlet beak. Three other tern species breed in the UK: sandwich (largest), little (smallest) and roseate (rarest). The first two have stubbier tails and non-red beaks and legs. All are summer migrants, generally most visible on passage: April/May or July/August. Likelihood of seeing them can be reckoned from their approximate 2019 British breeding numbers: Arctic (50,000 pairs), common (12,000), sandwich (12,000), little (2,000), roseate (100).

Of all the world's terns, it is the far-ranging Arctic that captures the imagination. In Britain, it favours northern island breeding sites, which it defends against all comers,

with the harsh fury of its cry as well as its stabbing beak. Considering its beauty and romantic wanderlust, it has been neglected by poets.

The nineteenth-century American Bret Harte's 'To a Sea-Bird' is unspecific, but seems best to fit the Arctic tern:

Little thou hast, old friend, that'snew;
    Storms and wrecks are old thingsto thee;
Sick am I of these changes too;
Little to care for, little to rue. —
    I on the shore, and thou on the sea.

All of thy wanderings, far and near,
    Bring thee at last to shore and me;
All of my journeyings end them here,
This our tether must be our cheer, —
    I on the shore, and thou on the sea.

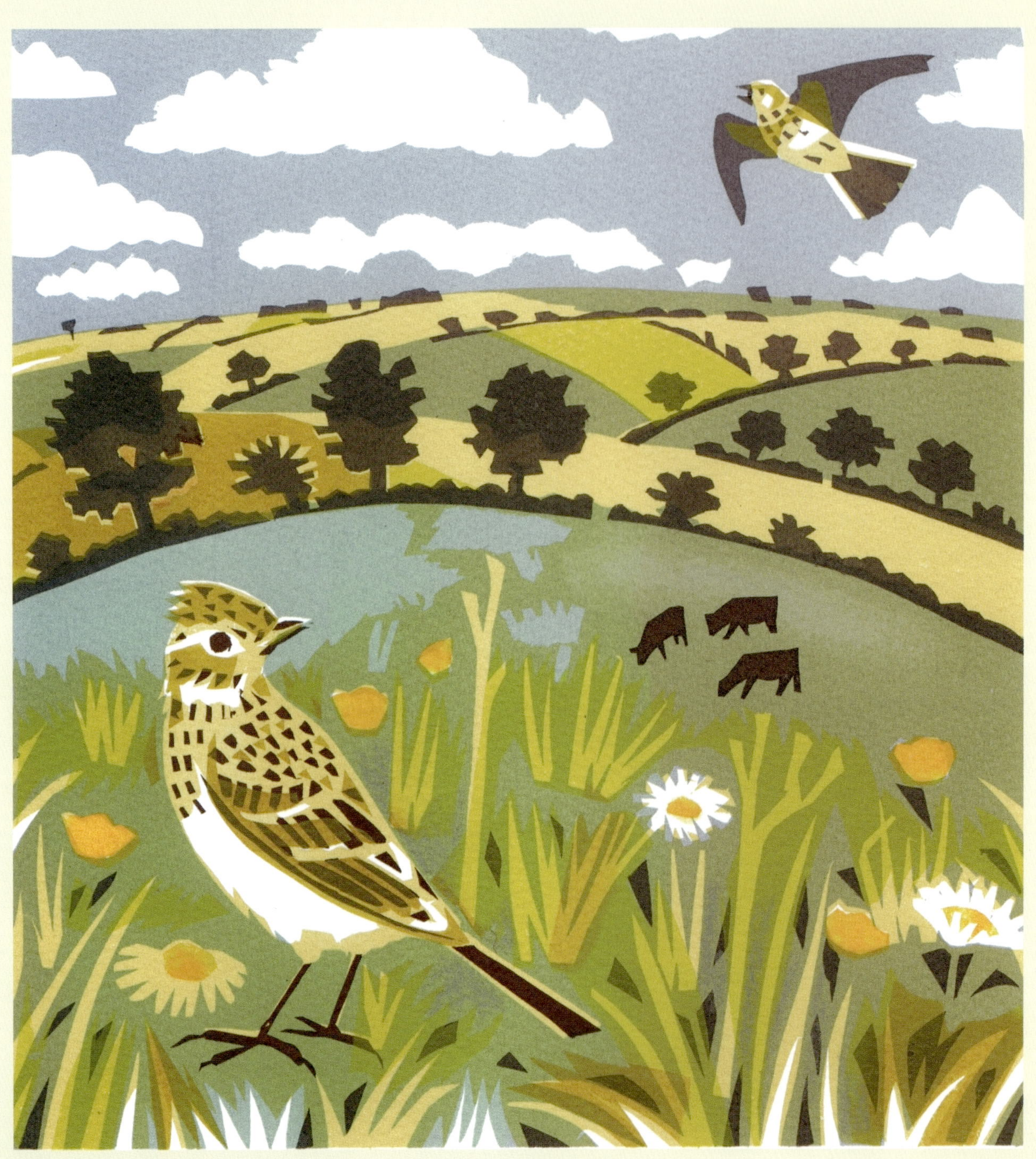

THE SKYLARK

Hail to thee, blithe Spirit!
Bird thou never wert,
That from Heaven, or near it,
Pourest thy full heart
In profuse strains of unpremeditated art.
[…]
Better than all measures
Of delightful sound,
Better than all treasures
That in books are found,
Thy skill to poet were, thou scorner of the ground!

Teach me half the gladness
That thy brain must know,
Such harmonious madness
From my lips would flow
The world should listen then, as I am listening now.

PERCY BYSSHE SHELLEY, from 'To a Skylark'

'What larks, what larks!' – as Joe says to Pip in Charles Dickens's *Great Expectations* – does indeed derive from the merry song of the skylark (*Alauda arvensis*); a lark, meaning a frolic, dating from the early nineteenth century and first recorded by Byron before being immortalised by Dickens. In 1813 Byron implored the poet Thomas Moore to join him and 'take, what in flash dialect, is poetically termed a "lark"'. In the Royal Navy, the captain can order the ship's company 'Hands to Dance and Skylark' to raise morale.

In English only the nightingale has inspired more poetry; and, like the nightingale, it is the skylark's song which has captured the imagination. It is one of the few birds that continues to sing in and beyond late summer. Farming brought it to our skies and – until legal protection in 1931 – to our tables. Now it is the subsidised intensive farmer who most threatens its existence. In 2018 there were still a million or so skylark pairs in Britain (a dozen pairs in London's Richmond Park), but that represents 10 per cent of its numbers up till the 1980s. In France and elsewhere it continues to be legally hunted.

Skylarks have always been considered a delicacy, not least for that inconceivable dish, 'lark's tongue'. 'Take 1,000 larks', one recipe begins. It should not be forgotten that, 'Alouette, gentille alouette' the charming nursery song, is a detailed account of plucking (*Alouette, je te plumerai*) a skylark for the pot. Mrs Beeton has a skylark recipe; and at the dinner celebrating the 1890 opening of the Forth railway bridge, one course was lark pie. At that date up to 40,000 of the dead birds were delivered daily, 'truck after truck on the Great Eastern', to the City of London's Leadenhall market (Paul Donald, *The Skylark*). In winter, when fatter, they were especially popular. If prices fell, the best-conditioned males (only males sing) were sold as caged songbirds; although it is unimaginable to contemplate the song, which can contain 460 syllables, separated from the ecstatic flight to such heights (2,000 ft/610m) the bird becomes 'day's black star' (W. H. Davies, 'Day's Black Star'). To watch an ascent is to witness one of the world's wonders, as the poets attest.

Shakespeare's song 'Hark, hark! the lark at heaven's gate sings' (*Cymbeline*, Act 2, Scene 3) and Shelley's '*Hail to thee, blithe spirit, Bird thou never wert*' beat all other openings; but Vaughan Williams's *The Lark Ascending* for violin and orchestra, inspired by George Meredith's 122-line eponymous poem, is the most popular artistic tribute. The composer selected 12 of its lines for the flyleaf of the published score, which was written in 1914 and first publicly performed in 1920. These are six of them:

For singing till his heaven fills,
'Tis love of earth that he instils,
And ever winging up and up,
Our valley is his golden cup
And he the wine which overflows
To lift us with him as he goes.

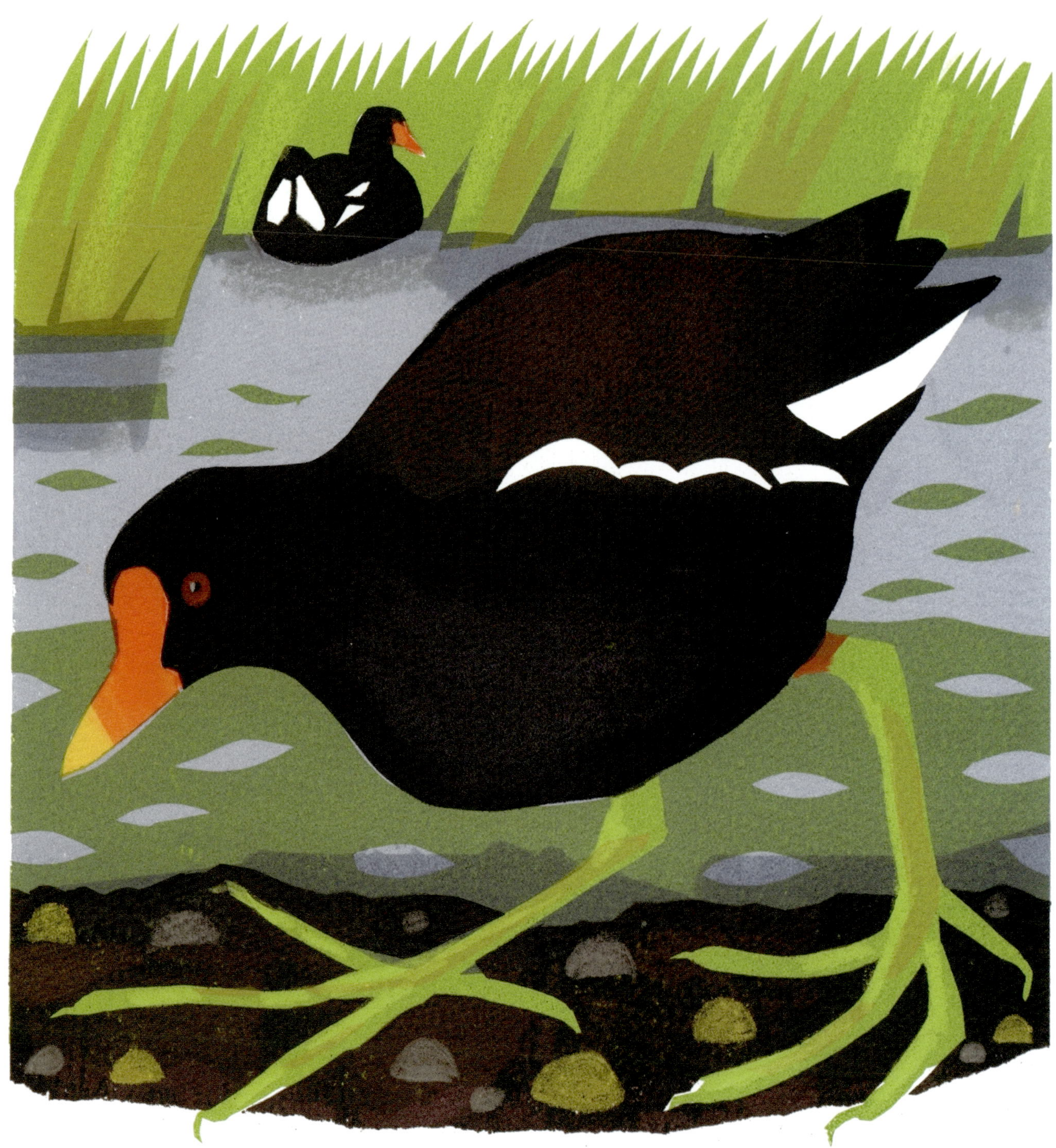

THE MOORHEN

'The most delicious form of idling imaginable', wrote 'BB' of coarse fishing in *Fisherman's Folly*. 'I know of no other sport, save wildfowling, in which you can withdraw from the workaday world so successfully, where all worries can be forgotten, where Nature can hold you so mesmerised, so entranced.' From Izaak Walton to the present, from England to China, the truth holds. In China, when a man is troubled, traditionally he takes up his pole and goes fishing.

Inevitably the moorhen (*Gallinula chloropus*) soon makes an appearance in *Fisherman's Folly*: 'The wheeling ripples from a questing moorhen which breaks up the reflection of the overhanging willow, the low summer hum of insects in the treetops, the motionless globes of the yellow water-lilies' – each inseparable and equally delightful. 'I am sure that a white spot on the tail of an animal or a bird acts as a guiding mark for their young. The moorhens possess it, and after that distinctive beacon their dusky babies scramble as the mother bird threads the sombre jungle of the reeds.' And, because carp fishing is also nocturnal: 'I hear the merry 'cruiks' of the moorhens, who seem just as busy at night as during the day.'

The moorhen has several colloquial names, including water hen, although its aggressive coot cousins are the more hen-like as they crowd to be fed. Appearances can deceive, as these descriptions from West Sussex in March 2019 by Tizzie Wells testify:

> There is a battle royal going on in my pond between a pair of mallard and moorhens. They are both trying to build nests. If they cannot co-habit happily, I would like the ducks to triumph. The moorhens are more aggressive, but the ducks bigger, and the moorhens slightly frightened of them.

A week saw the problem resolved.

> I'm afraid the moorhens have triumphed, and the ducks have gone. The former must have irritated the ducks beyond endurance – they can't have actually frightened them. The moorhens are looking very perky and making a nest themselves where the ducks had started to make one before they decamped. Much the most sheltered and sensible place.

The moorhen is indeed 'watchful'. In a field it runs for cover like a rabbit, habitually loath to fly. Surprising then that the British population is trebled to a million in winter by migrants from Scandinavia. Surprising too that the moorhen is still classified as a game bird, with a shooting season from 1 September to 1 February. The Cypriot community in north London was once keen on the meat but, as a Kentish Town butcher lamented recently, 'the young don't eat game'; and he meant the standard fare of pheasants, woodpigeons, etc. Oscar Wilde's quip – 'Nature is what we call birds flying uncooked' – no longer applies, even in Mediterranean countries noted for their cuisines, their appetites and open seasons for songbirds are in the twenty-first century tempered by European law.

Moorhen eggs were once a worthy substitute for those of green plover (lapwing), now banned in Britain, and black-headed gulls, a licensed delicacy. Often the nest is precariously placed on the farthest branch of a half-submerged waterside tree where, as John Clare wrote in 'The Moorhens Nest':

> *– though danger comes*
> *It dares and tries and cannot reach their homes*

Not today. Mink, whose wild status in Britain is owed to escapees from fur-harvesting mink farms, are especially devastating (see The Kingfisher, page 99), even rendering the birds extinct in the Western Isles. The moorhen is notably defensive and solicitous of its young. Throw food and the parents will feed the chicks before themselves.

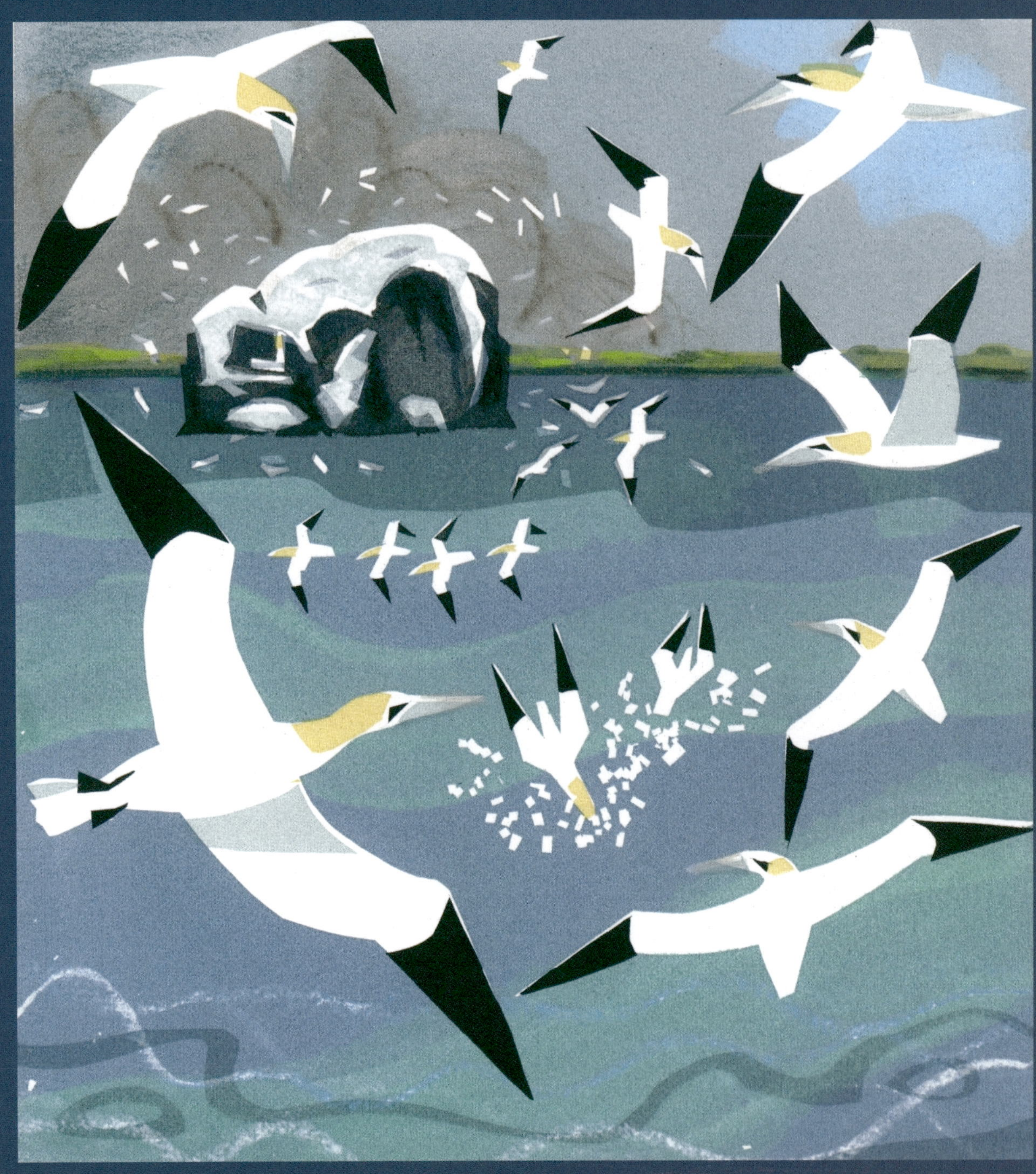

THE GANNET

Becalmed off Ailsa Craig in the Firth of Clyde, the air filled with the grating din of breeding gannets, I asked Freddy Hesketh, hopeful young fisherman, what they were saying. '"More fish"', he said. But where were they finding their favourite mackerel and even scarcer herring? Our party had caught nothing. Fish stocks are reported to have improved; nonetheless gannets have long adapted to human profligacy and supplemented their fish diet by scavenging. It was in 1947 that the bird was first recorded filching a piece of bread.

The gannet (*Morus bassanus*, after the Bass Rock (illustrated) in the Firth of Forth) is Britain's largest seabird, with a 6ft (1.8m) wingspan and weighing up to 8 lbs (3.6kg). Adaptability has dramatically increased its numbers. The latest surveys (published 2015) show the Bass Rock has recaptured its position as the northern gannet's world HQ (75,000 pairs, up 24 per cent since 2009), overtaking St Kilda (60,000 pairs). The Bass shines white in summer from the birds. Grassholm, off Pembrokeshire, is the largest colony in Britain excluding Scotland. Satellite telemetry has shown Bass-based gannets have foraged a maximum 335 miles (540km), the farthest point 144 miles (232km), to feed their young, regurgitating the catch on their return.

…a barrage of gannets
whose detonations
explode the green into white.

NORMAN MacCAIG, from 'Rhu Mor'

Even in winter, when many migrate south to the Bay of Biscay and off West Africa, enough remain to make sightings commonplace.

In L. A. G. Strong's 1932 novel, *The Brothers*, an 'informer' is executed by being floated upright in a sea loch with a herring fixed to his head. Soon a 'solan goose' (from Norse, *sula* = cleft stick, describing the gannet's black wing-tips, when crossed) appears as executioner. Afterwards 'they could not separate the goose from the informer. The iron beak had split his head as a wedge splits a piece of wood'. A spongy bone plate at the base of the diagrammatically lined bill equips a gannet to hit the sea at 60 mph. On average they penetrate the water to a depth of 15 feet (4.6m). Christopher Rush (*Hellfire and Herring*) recalled from his post-Second World War Fife childhood: 'The oldest fishermen swore they could smell a shoal of herring…as accurately as a gannet, diving from two hundred feet into the waves could detect the glimmer of a fish another two hundred feet beneath the surface.' Gannets can live for 25 years.

They used to be a familiar food. Daniel Defoe described the meat as 'rank'. London restaurants offered gannet as 'Highland goose' in the Second World War. Locals are still licensed to fillet and kipper the gugas (nestlings) from Sula Sgeir, off Lewis. Into the 1920s, St Kilda's cragsmen defied death to take gugas from Stac an Armin, at 627 feet (191m) Britain's tallest sea rock, the same height as London's BT Tower. Today, with St Kildans long gone, the gannets are undisturbed.

Gannets fall like heads of tridents,
bombarding the green silk water
off Rhu Mor…

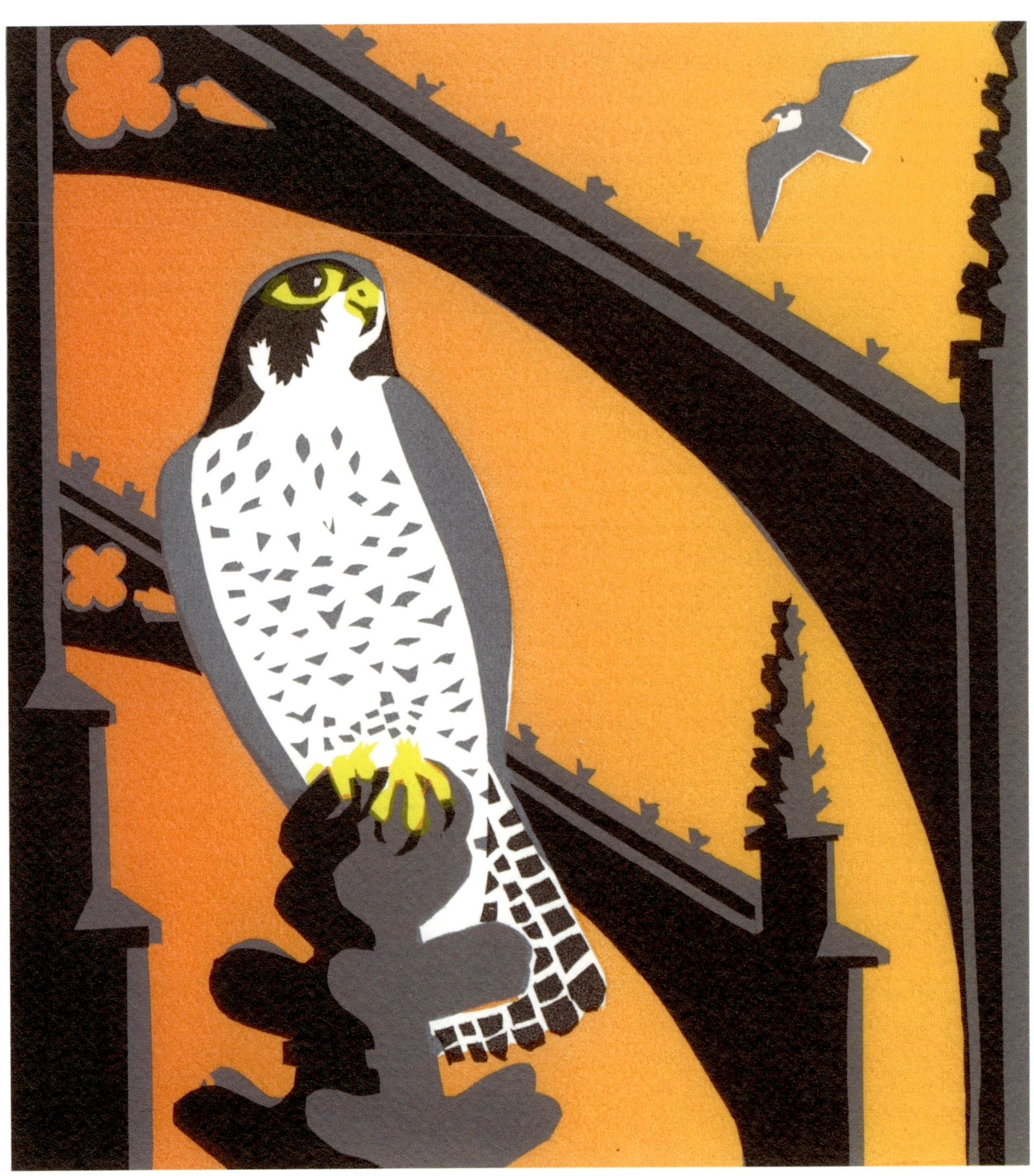

THE PEREGRINE

The peregrine (*Falco peregrinus*, wandering) is the world's fastest creature, revered for millennia, subject of tribal worship, essential to Europe's chivalric tradition. As 'peregrination' implies, the peregrine exemplifies freedom. Its supreme chronicler, J. A. Baker, wrote: 'You cannot know what freedom means till you have seen a peregrine…roam at will through all the far provinces of light.' (*The Peregrine*).

From the late 1950s Baker, wage-work permitting, followed the bird for a decade with bicycle and binoculars in coastal Essex. His book is a journal from autumn to spring, years and landmarks unspecified. The bird was facing extinction. The book is a threnody to its beauty in expectation of its extinction – 'burnt away by the filthy, insidious pollen of farm chemicals'. Banning DDT initiated the peregrine's revival. From 300 UK pairs (1973) there are now 1,500 – 30 of them in London, probably the highest density apart from New York. They are at their safest and most cosseted in cities. Thanks to the government's wildlife watch, Natural England, and the London Peregrine Partnership, the Battersea Power Station Development Company spent more than £100,000 re-siting its resident peregrine pair. Cathedrals are a favourite urban location in cities outside London: Lincoln, Norwich, Canterbury, Chichester, Salisbury all have peregrines. Tate Modern is Britain's most visited peregrine home. They hunt at dawn and spend the day watching the world go by. Research has revealed city lights can make them night hunters, killers of nocturnal travellers like elusive water rails and corncrakes. Pigeons are their favourite prey, but they can kill bigger birds, even the buzzard. The Battersea birds' most surprising kill was a hen pheasant. London peregrines are now partial to the rose-ringed parakeet.

John Haddington, a modern Merlin, let rip a box-kite at Mellerstain in Berwickshire which reached such a height it magicked a curious peregrine from the Cheviots 15 miles to the south. Was this a passing bird or had it seen the kite from its Cheviot eyrie and come to investigate an intruder?

In the ancient hierarchy of falconry, originally a Chinese sport and still popular in several countries (hence a goad for criminal egg-smuggling), the peregrine was fit for a prince, a gyrfalcon for a king, an eagle for the emperor. Juliana Berners's 1486 treatise on manners (including falconry), *The Boke of St Albans*, is probably the first printed book in English by a woman and out-sold all others until J. K. Rowling's *Harry Potter* series. Falconry terms include 'fed up', 'under his/her thumb', 'haggard', 'hoodwink', 'bated breath', 'cadge', 'mews' and 'boozer'.

As Carry Akroyd says, 'you are much more likely to see peregrines sitting around than doing their speedy stuff'. And speedy it is, a level flight average of 60 mph, a timed stoop of 217 mph – various mechanisms are triggered to shield its eyes from the blast. The artist Lawrence Preece was working in his steeply sloped Ffestiniog garden when he heard 'a genuinely frightening roar of an approaching jet fighter'. It was a peregrine, which hurtled passed him in mid-stoop focused on some bird farther down the hill.

June is a good month for seeing them, when the young need feeding or are taking their maiden flights. The male 'tiercel' – so called because it is a third the size of the female 'hawk' – assumes brooding duties from the third day after the hatch. Fledged birds are brown and buff into their second year.

It was their flight that transported Baker:

Bending over in a splendid arc, she plunged to earth…
I saw fields flash up behind her; then she was gone
beyond elms and hedges and…I was left with nothing
but the wind.

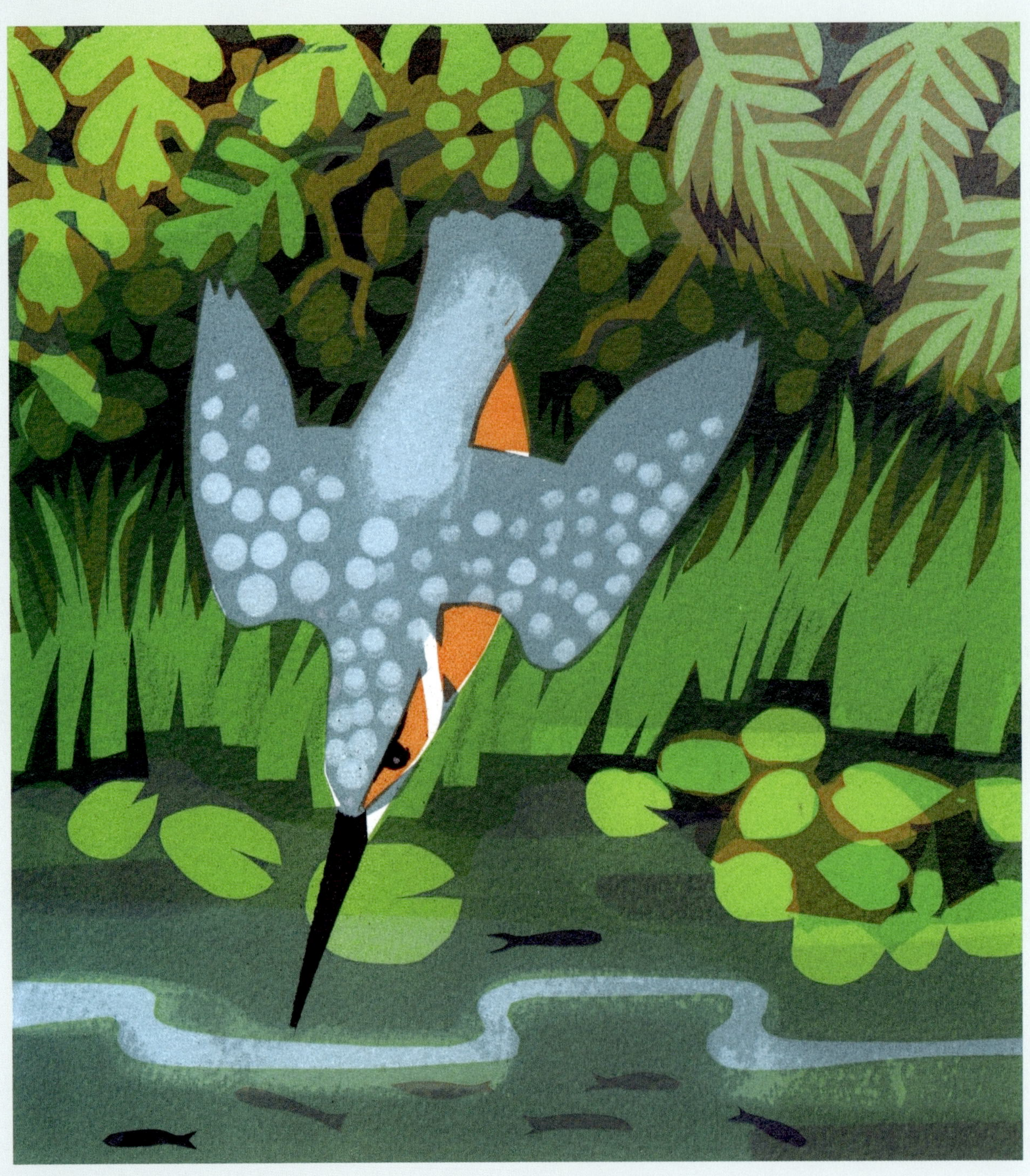

THE KINGFISHER

The kingfisher (*Alcedo atthis*) derives its 'halcyon' (*alcedo*) name from the Greek goddess, Halcyone or Alcyone, calmer of storms. In one legend Halcyone drowned, but the compassionate gods turned her into the beautiful bird. In another legend the waves were still during the seven days before and after the winter solstice, allowing the kingfisher to nest on the sea. Hence 'halcyon days', tranquil weather; and the link of the dazzling blue plumage with halcyon summer or, for Shakespeare, halcyon Indian summer: 'Expect Saint Martin's [11 November] summer, halcyon days' (*Henry IV*, Part 1). In the Middle Ages, the bird's corpse was thought to be endowed with beneficial properties, which may have made it emblematic of the 'vital spark' of life or soul that survives death. It is certainly as a vital spark that the kingfisher is most commonly seen; flashing past on a dead-straight course, sounding an appropriately urgent 'beep beep'.

So when the shadows laid asleep
From underneath these banks do creep,
And on the river as it flows,
With eben shuts begin to close;
The modest halcyon comes in sight,
Flying betwixt the day and night;
And such an horror calm and dumb,
Admiring Nature does benumb.

The viscous air, wheres'e'er she fly,
Follows and sucks her azure dye;
The jellying stream compacts below,
If it might fix her shadow so;
The stupid fishes hang, as plain
As flies in crystal overta'en,
And men the silent scene assist,
Charmed with the sapphire-winged mist.

ANDREW MARVELL, from 'Upon Appleton House'

It is the azure band of feathers down its back which especially dazzles. In fact, the kingfisher's amazing colour does not come from pigment but light, as it strikes modified layers of cells in the feathers. Glimpsed in flight – or, if you are lucky, seen perched or diving – one is reminded of its astonishing brilliance; although, breast to the fore, it can appear rust brown. Sight of them is particularly memorable because invariably it is a surprise – such as on Hampstead Heath ponds, where they still survive.

They are well-named king of fishers, a pair feeding a brood will catch over a hundred fish a day. To nest they burrow 1½ feet (46cm) into pond or river banks, accumulated fish remains causing a notorious stench. As 'BB' wrote in his children's book classic *The Little Grey Men*: 'Kingfishers are filthy birds in their nesting habits, and it was always a source of utmost amazement that such gorgeous and kingly beings could be so dirty.' Burrows make them prey to hole-living predators, in modern Britain most devastatingly the mink.

Ice is their deadliest enemy. In a prolonged freeze, they may take to the seashore in search of fishable water. It was reckoned only 10 per cent survived the winter of 1962/63; but they are resilient, capable of multiple broods of up to 10 to restore numbers.

And then, from wastes of stub and nothing came
The Kingfisher, whose instancy laid bare
His proof that ice and sapphire conjure flame.

PETER SCUPHAM, from 'Kingfisher'

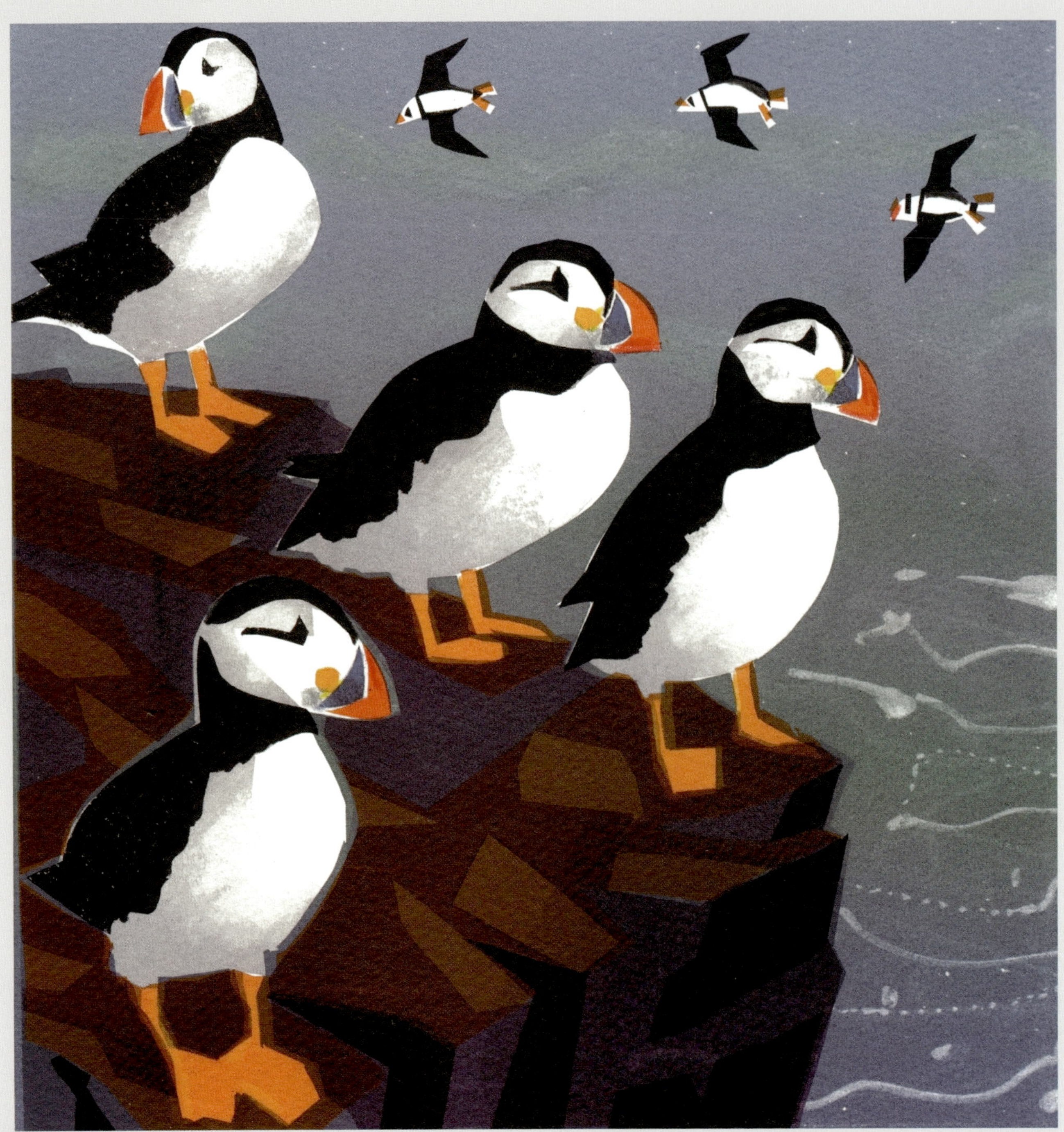

THE PUFFIN

The cutest auk, the puffin (*Fratercula* ('little friar') *arctica*) is the only seabird in the top-10 most popular UK birds list. Puffin children's books, launched in 1940, prevail; and Lundy Island's puffin stamps – *lunde*, the bird's Norse name – for which puffinage is charged, is the world's oldest private postal service. Forty thousand items are posted annually. The island is leased by The Landmark Trust, that most civilised conservation body founded by the late Sir John Smith.

The collective noun is a circus of puffins. The carnival-mask beak acts as a spade for digging the nest burrow and can hold and catch fish simultaneously, 61 sand eels and two rocklings the record. To feed itself and young requires 450 sand eels a day, which may require two-minute dives to depths of 220 feet/67m (*The Seabird's Cry*). A geo-locator revealed one Shetland-based bird had to fly 248 miles to bring back fish. The beak's brightness depends on carotenoid-rich fish. The brighter the beak, the healthier the puffin. In winter, when puffins roam the seas, the beak's bright carapace is shed, and they are nondescript.

From March through August puffins are ashore. They can live over 30 years and nest in warrens, on occasion co-existing with rabbits, the turf so laced with their burrows it sometimes collapses. A single egg is laid in May. Incubation takes six weeks, and for a further six the puffling is fed underground – protected from predatory gulls but not rats, the scourge of seabird colonies. It is then abandoned. Goaded by hunger, it walks down to the sea in nocturnal safety and does not return for two years.

Scotland's seabird colonies, home to 45 per cent of the European population, are in decline, with puffins verging on the endangered list. Warmer seas are blamed, affecting the plankton on which sand eels depend. Grey seals are partial to puffins, but their button-eyed white pups melt the hardest heart. It is indicative that the Grey Seal Protection Act 1914 (except in Ireland) was the world's first on behalf of a mammal; and that in 1977 there was a public outcry after an emergency cull in Scotland. A century on there are 200,000 grey seals in the British Isles – 50 per cent of the global population – and 40 per cent of the British population are in Scotland. Even a cull to protect endangered puffin colonies remains a political non-starter. St Kilda has the UK's largest colony but is shrinking. By contrast on the Shiants, where a £900,000 extermination scheme (half EU-funded) has rid the islands of its unique black rats, the population is stable. In 2017 the Isle of May had its best breeding season for 30 years, with 90 per cent success. Carry Akroyd writes of May: 'We used to see the puffins lined up along the clifftop looking out to sea and called it "going past the bus stop". One morning, there were none. We went over to the other side of the island and there they all were, looking the other way because the wind had changed.'

Papal dispensation once allowed puffins to count as fish in Lent. They are still eaten in Iceland and elsewhere. David Profumo found their gamey red meat preferable to guillemot (*Country Life*, 12/4/17) but discovered day trippers do not qualify for a drink at St Kilda's Puff-Inn. There is also Puffin's, the private luncheon club, founded by the late Sir Iain Moncreiffe of that Ilk, Bart, CVO, QC, Albany Herald, 24th Chief of the Name and Arms of Moncreiffe. It was named after his first wife, Lady Diana Denyse May, 23rd Countess of Erroll in her own right and Lord High Constable of Scotland, whose nickname was Puffin. Membership is by invitation, congeniality the sole requirement. Their sons, Merlin Hay, 24th Earl of Erroll, who has the proprietorial role of his father, and fellow member Peregrine Moncreiffe of that Ilk, 25th Chief of the Name and Arms of Moncreiffe, safeguard the club's traditions.

Alexander Hesketh of immortal Hesketh Racing fame, remembers Sir Iain, on a visit to Aigues-Mortes in southern France, swimming in the Mediterranean wearing 'a brown trilby as if at Perth Races'.

A puffin was found wandering in Sloane Square in September 1984.

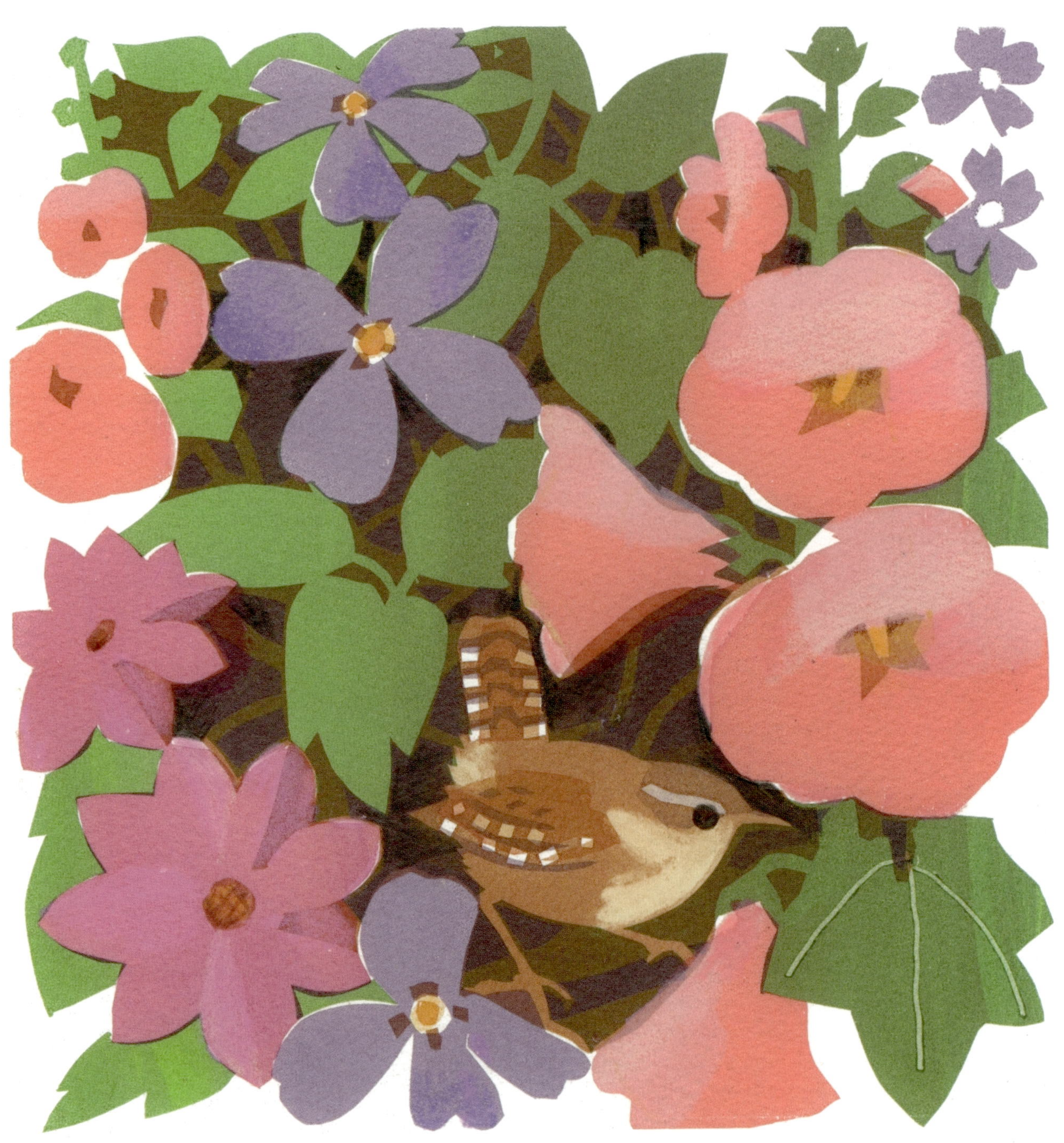

THE WREN

The charm of the wren (*Troglodytes troglodytes*, 'cave dweller') has earned its 'Jenny' nickname and made it the emblem on the UK's prettiest coin, the Victorian-designed farthing (withdrawn 1960).

One July morning I was greeted by a barrage of wren song, and this in the treeless depths of London's Olympia. At that otherwise song-less time of the year, it seemed to be a newly fledged brood finding their voices; a forceful reminder that it is our most ubiquitous bird, as at home in the concrete jungle as the wilderness, and it has the most powerful song of diminutive songsters – per unit weight, 10 times stronger than a cockerel's crow. Moreover, it can sing throughout the year, unlike most songsters, and females sing a little too. The exuberant song is in keeping with its jaunty upturned tail: a torrent of lisped riffs with a linking brrrrrr, loud enough to pierce the roar of traffic or the sea. Like some other birds (robins, blackcaps) its alarm call is a time bomb, 'tickety-tick-tick'.

Mousy by nature, it is more often heard than seen, hence its 'cave dweller' name. All habitats have its favoured clefts and crevices, where it also finds its favourite food, spiders. This adaptability explains its presence on far-flung St Kilda, where long isolation has created an endemic subspecies, the St Kilda wren (*Troglodytes hirtensis,* after Hirta, largest of St Kilda's four islands), a paler version of the russet mainland bird. Its fame has eclipsed other island sub-species on Shetland, Fair Isle and the Outer Hebrides.

Wrens are famous and eccentric nest builders.

There was an Old Man with a beard,
Who said, 'It is just as I feared! —
Two Owls and a Hen, four Larks and a Wren
Have all built their nests in my beard.'

EDWARD LEAR, from *A Book of Nonsense*

Wrens have built their domed nests virtually everywhere, from inside a skull to the running board of a working lorry; so why not an unkempt beard? An odd characteristic is the insurance of building false nests. The average is six, despite a complexity surpassed only by that of the long-tailed tit. The female's choice is then softly lined. Hence the name 'cocks' nests' for the unlined rejects. There are two broods of up to 10 eggs, the young fed 50 per cent on caterpillars. Such fecundity sustains the population even after the hardest winter.

It's usually beeches, with their shallower root
systems. One night there's a storm, and the tree
creaks hugely and lies down at last, after seventy or
eighty or ninety years of being vertical, and all the
tracery of roots, still clotted with earth, presents its
secret intricacy. Subterranean much is a sudden cliff-
face; the air gets in at it, and the light. Keep a wry eye
out for a wren's nest. Bits of dried fern, and leaves,
and blades of pale grass, all woven among the roots,
not like the cup of the conventional nest, but a little
globe, the tree roots forming the back wall, with an
entrance hole, slightly wider than a ten pence piece,
at the front. Then up to eight eggs, a little bigger
peas, white, minutely constellated with blood,
for the bird to warm in her mossy cave while out
around and above the high wood lurches its crown
and creeks. Eight dark tiny embryos, each rising in
its shell oven. Then a cave of gapes, the death of a
hundred thousand spiders. Feathers and attentive
eyes, crammed in the hot slum. Approach now and
they stream out, one after the other, in a tracer fire of
brown feathers, to skulk among the ticking foliage.
And here's the nest, in November, the empty
cave without a fire..

GERRY CAMBRIDGE, 'Troglodytes troglodytes'

In glacial 1963 four-fifths of the 20 million UK population (under half that, 2019) died. A 'hot slum' is the answer. Sixty birds in a nest box is the record (*Birds Britannica*), death from suffocation a hazard.

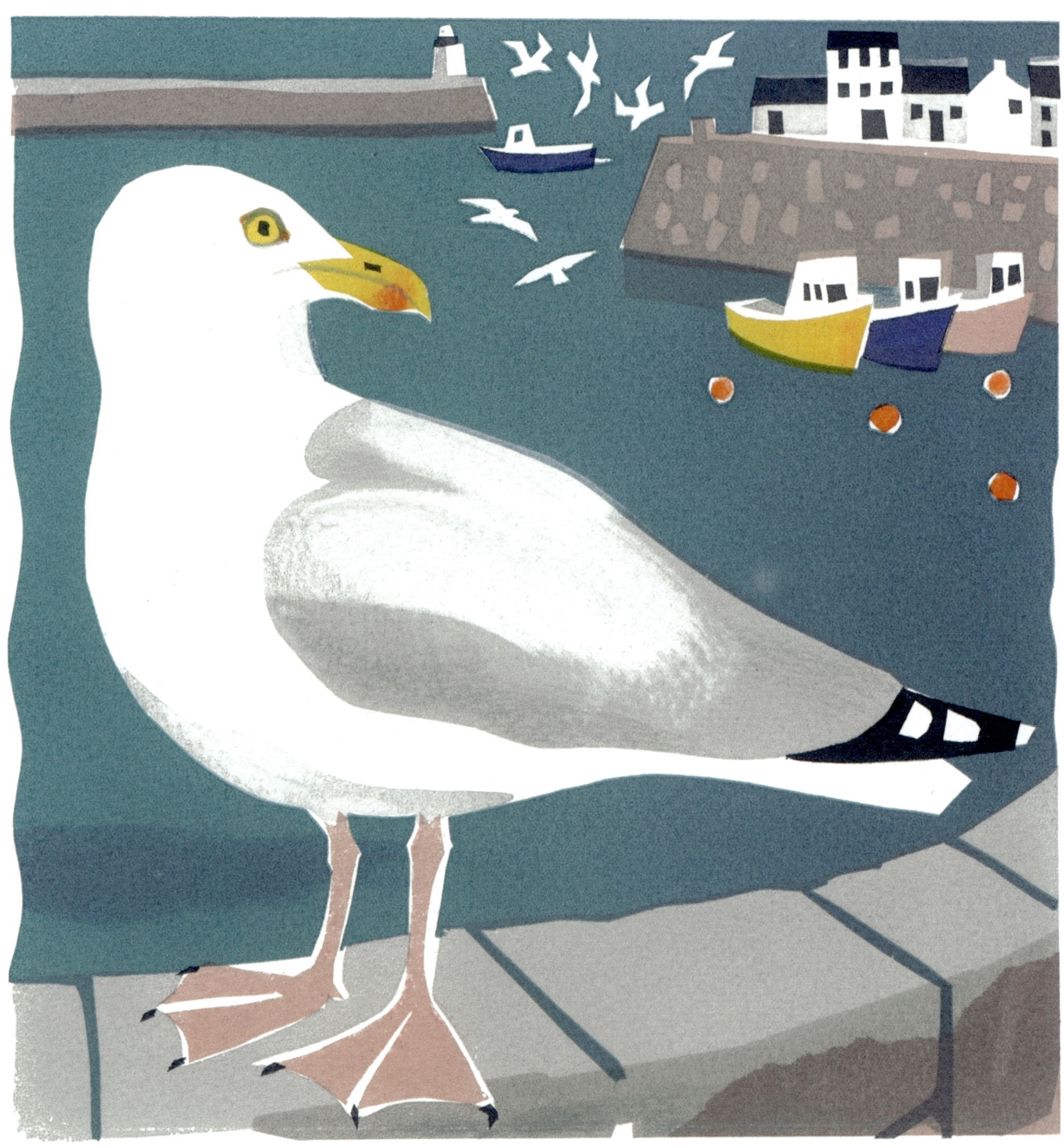

THE HERRING GULL

'Seagull' usually means the herring gull (*Larus aregentatus*). Until the latter part of the twentieth century they were indeed largely confined to the seaports. The artist John Bellany, son of a sea captain, was first allowed on a fishing boat when he was 13: 'It was totally different to what I'd dreamed. For one thing, I had no idea the seagulls would be so gigantic. When they spread their wings, they're almost the size of a man, and when the net comes in, they're right on top of you, making this deafening noise' (John McEwen, *Bellany*). He was describing herring gulls, which have wingspans of up to 5¼ feet (1.6m). These were the days when fish were still plentiful, and there was no need to catch the lowly pollock or prawns. In 2019 the latter constitute Britain's main fish export but were traditionally chucked overboard as net-entangling 'vermin'.

Lack of fish has hastened the herring's decline and its urban inclination, sometimes amounting to harassment. Their piratic behaviour hogged 2015's silly-season headlines: 'Seagull attack cancels concert'; 'Hannigull Lecter'; 'Sexist Beasts! Vicious Gulls Target Women'. Cartoonists followed suit. Pugh showed a husband complaining to his wife: 'Feeding seagulls is one thing, but do they have to join us for lunch?' Bob's 'Secret Weapon' had a flight of herring gulls in airman's goggles defending the realm. Even prime minister David Cameron pronounced: 'I think a big conversation needs to happen about this.' A gull had snatched his seaside-holiday ham sandwich but, magnanimously, he did not hold that 'against the entire seagull population'.

Cornwall, the UK's Riviera, provided most reported trouble. In 2014 Lynda Charlton, an artist, was 'jumped' by several herring gulls while eating a sandwich. 'This was no ordinary attack by a hungry gull,' she said. 'They seemed organised and working as a team.' Such reports encouraged visions of Tippi Hedren in Hitchcock's *The Birds*, distraught in a gull-besieged phone box. Women seemed particularly vulnerable. 'They've learnt that girls are more likely to drop their food and run for it,' said Becca and Ash,

Cornish lifeguards. One girl had her iPhone snatched and dropped out to sea. Cornish horror stories concerned the gull-inflicted deaths of Roo, a Yorkshire terrier, and Stig, a 20-year-old tortoise. It was not only Cornwall. In Hyde Park, a rogue gull repeatedly drowned feral pigeons in the Serpentine.

'They've learnt to live with us…now they're getting into conflict with us, too,' said Keith Bretton of the British Trust for Ornithology. The RSPB advises the avoidance of nesting gulls, especially the 100,000 pairs of various urban-breeding species, and to carry an umbrella. Gulls can projectile excrete what they have eaten within minutes. The *Cornishman* suggested a laser gun as a deterrent; a parent, a pump-action water pistol.

Conflict has been encouraged by the easy food offered by the proliferation of landfill dumps, to such a degree the herring could be renamed the garbage gull. With recycling depots replacing dumps, the gulls are driven to hunt for scraps like pigeons. The phenomenon can be exaggerated. In 2016 four satellite-tagged herring gulls nesting on a St Ives cinema roof provided a case for the defence. Three of them spent most of their time at sea – two of them up to 18 miles (29km) offshore – or followed inland ploughs.

Yet, contrary to noisy appearance, the native population – boosted to five times its resident UK 250,000 by winter migrants – has halved since 1970 and, such is the conservation concern, it is now red-listed, although its nest and eggs can still be legally destroyed. All gulls are natty – super-clean, with never a ruffled feather. Decline alerts us to the herring's mature beauty, fully attained only after a second winter: the red-flecked golden blade of a beak, smooth-white breast, grey-mantled wings with their domino pinions; and that majestic wingspan, tilting mastery of the breeze and keening oceanic cry, which by popular demand still introduces BBC Radio 4's *Desert Island Discs*.

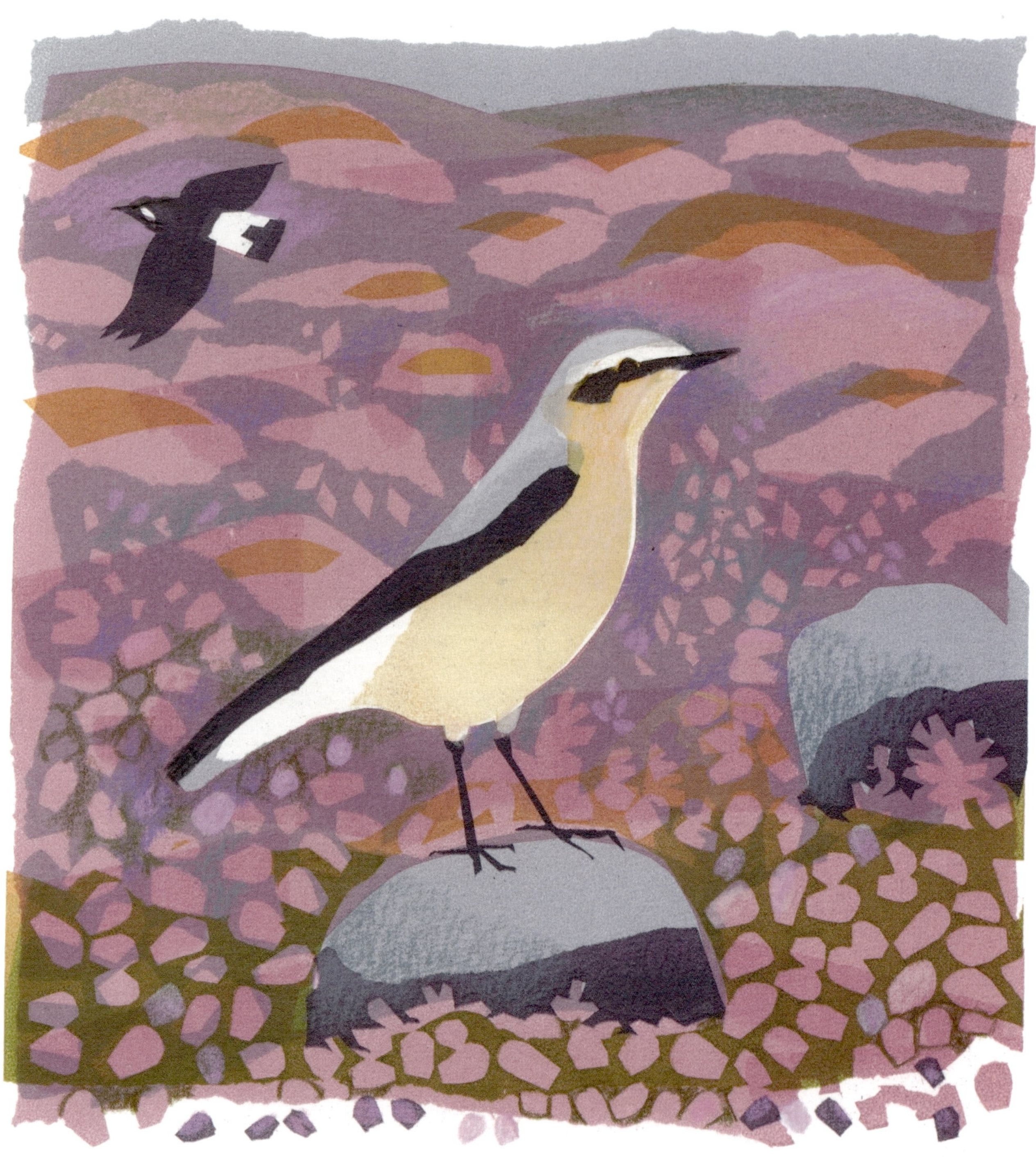

THE WHEATEAR

'To those who are attracted to solitary, desert places, who find in wildness a charm superior to all others, the wheatear [*Oenanthe oenanthe*], conspicuous in black and white and bluish grey plumage, is a familiar figure – a pretty little wild friend; for he, too, prefers the uncultivated wastes, the vast downs, the mountain slopes, and the stony barren uplands', wrote W. H. Hudson.

Stoniness is the theme of Thomas A. Clark's untitled wheatear poem:

> *The little one of the stones*
> *The clachoran the wheatear*
> *Restless flitting bobbing*
> *I watch it in the heather*
> *Caught between impulses*
> *Lighting lifting hovering*
> *its note like two stones*
> *Struck lightly together*
>
> From *Tormentil and Bleached Bones*

The similarity of their alarm call to struck stones is what gave the chats – in Britain stonechat, whinchat and wheatear – their name, although they do also have a song.

Like the robin, the chats were classified as thrushes (*turdidae*) but are now considered flycatchers (*muscicapidae*). The white-rumped northern wheatear is a summer migrant. Their scientific name, *Oenanthe*, combines the Greek words for 'wine' (*oenos*) and 'flower' (*anthos*) and celebrates the bird's spring return to Greece when the grapevine blooms. This gives no indication of its astonishing migratory range from the northern regions of sub-Saharan Africa. Northern wheatears divide east and west, so today they nearly meet in Arctic Canada. Birds from the east can migrate 15,600 miles (25,105km); from the west, via Greenland, a mere 7,000 miles (11,265km), but this includes a transatlantic flight of up to 2,000 miles (3,219km). It may look chunkier than the swallow, but both these long-haul flyers weigh an ounce (28g).

Its former British abundance is testified by 90 vernacular names. 'Wheatear' is from Old English *whit* (white), *eeres*, *ers* (arse). Cocks can fan their black and white tails for long periods as a defensive distraction. Like the skylark, its numbers were devastated by the pot. Squire Wilson, royalist lord of the manor of Eastbourne, was saved by his wife treating a Cromwellian search party to wheatear pie, while he feverishly burnt papers upstairs. Charles II was particularly partial to wheatear.

The southern migration used to be mercilessly trapped on the Sussex Downs in August and September. One nineteenth-century trapper bagged more than 1,000 in a day. By Hudson's time, dwindling numbers were pricing the bird out of a market, but in 1900 he could still protest: 'It is not fair that it should be killed merely to enable London stockbrokers, sporting men and other gorgeous persons who visit the coast, accompanied by ladies with yellow hair, to feed every day on "ortolans" at the big Brighton hotels.' The wheatear was called the 'English ortolan' after the ortolan bunting, rarely seen in Britain. In France, wheatears are now mercifully spared but equally 'protected' ortolans are still at risk in southern Europe.

As a protected bird, the wheatear has long been off British menus; but down-land ploughing, intensive farming and diminished populations of grass-cropping rabbits and sheep have helped reduce the annual British influx to 240,000 pairs, now mostly confined in the nesting season to Wales, the Lake District and Scotland.

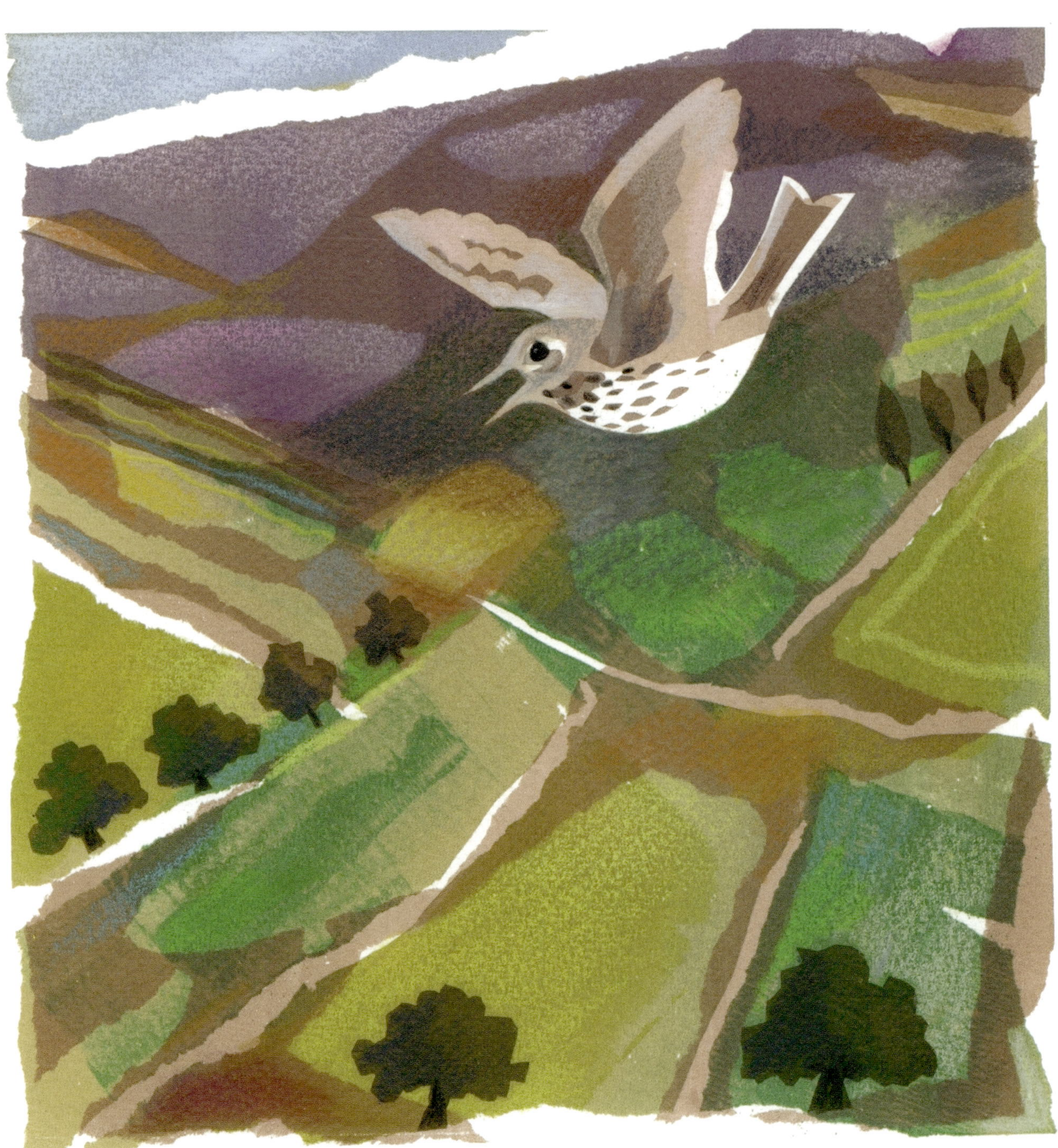

THE MEADOW PIPIT

Since the UK population is now predominantly urban, the summer holidays offer most of us the best chance of seeing a pipit. There are at least 35 species, three of which are common to Britain: meadow pipit (*Anthus pratensis*), rock pipit (*Anthus petrosus*) and tree pipit (*Anthus trivialis*). The three are virtually identical in appearance and song; habitat, as their names suggest, their most reliable distinction. There are 2 million UK meadow pipits, 40,000 rock pipits and 100,000 summer-migrant tree pipits, so 'a pipit' will invariably mean a meadow pipit. In the popular propagandist wartime film comedy, *Tawny Pipit*, that rare continental visitor was of necessity played by home-based meadow pipits, on the correct assumption that the audience would not spot the difference.

Pipits were first distinguished from tit (small) larks in 1795 to form a new genus *Anthus*; its onomatopoeic English name was accepted in 1833 and the three types differentiated. Today pipits are classified Motacillidae with wagtails. Pipits are ground nesters and share the same summer song flight, ascending lark-like to a small height, before floating ('parachuting') to earth or perch. The meadow pipit can share the others' habitat, but is the hardiest of the three; usually the only songbird seen above 2,000 feet (609m). It is a favourite host of the cuckoo, which accounts for the western Highlands and Islands now being prime cuckoo country, both species sustained by abundant caterpillars.

Meadow pipits were once ubiquitous, albeit preferring, as a nineteenth-century ornithologist wrote, 'moist situations, such as water-meadows, brook-sides, and turnip fields, after rain' or 'following cattle'. Today it is principally an upland bird (moss cheeper, Scotland; heather lintie, Cumbria), most widely spread in winter when it flocks the lowlands with migrants from northern Europe. A handful still breeds in Greater London at Wanstead Flats and Wormwood Scrubs. Autumn and spring migrations result in occasional sightings of large flocks, even in central London: 500 passed over Regent's Park on 5 October

2000; and 700, Hampstead Heath on 24 March 2006. Overall the UK population has halved since 1970.

Lord Grey and W. H. Hudson , who were close friends, were particularly fond of this easily disregarded bird, a favourite prey of raptors – even the golden eagle, as *Springwatch* revealed – dismissively filed by raptor enthusiasts as a 'pipit unit'.

> *Hawks could teach bullets a thing or two —*
> *see one precisely repeating*
> *the terrified unpredictable zigzags*
> *of a mountain pipit.*
> NORMAN MacCAIG, from 'Three Figures of Beethoven'

Hudson extolled its 'pretty winter plumage – the olive-browns and dull blacks, the whites and the cream faintly tinged with buff on the striped breast'; and the pleasure it gives 'of a pensive kind with something of a mystery in it'. He delighted equally in the 'bell-like tinkling' of its summer song and in winter its 'one sharp sorrowful little call, or complaint, the most anxious sound uttered by any small bird in these islands'. A sound suited to stony wastes and clouds flying 'when the thin but penetrative little sad call seems more appropriate than ever' as the escaping bird is flung 'a mere dead leaf or feather against the blast' (*Birds and Green Places*).

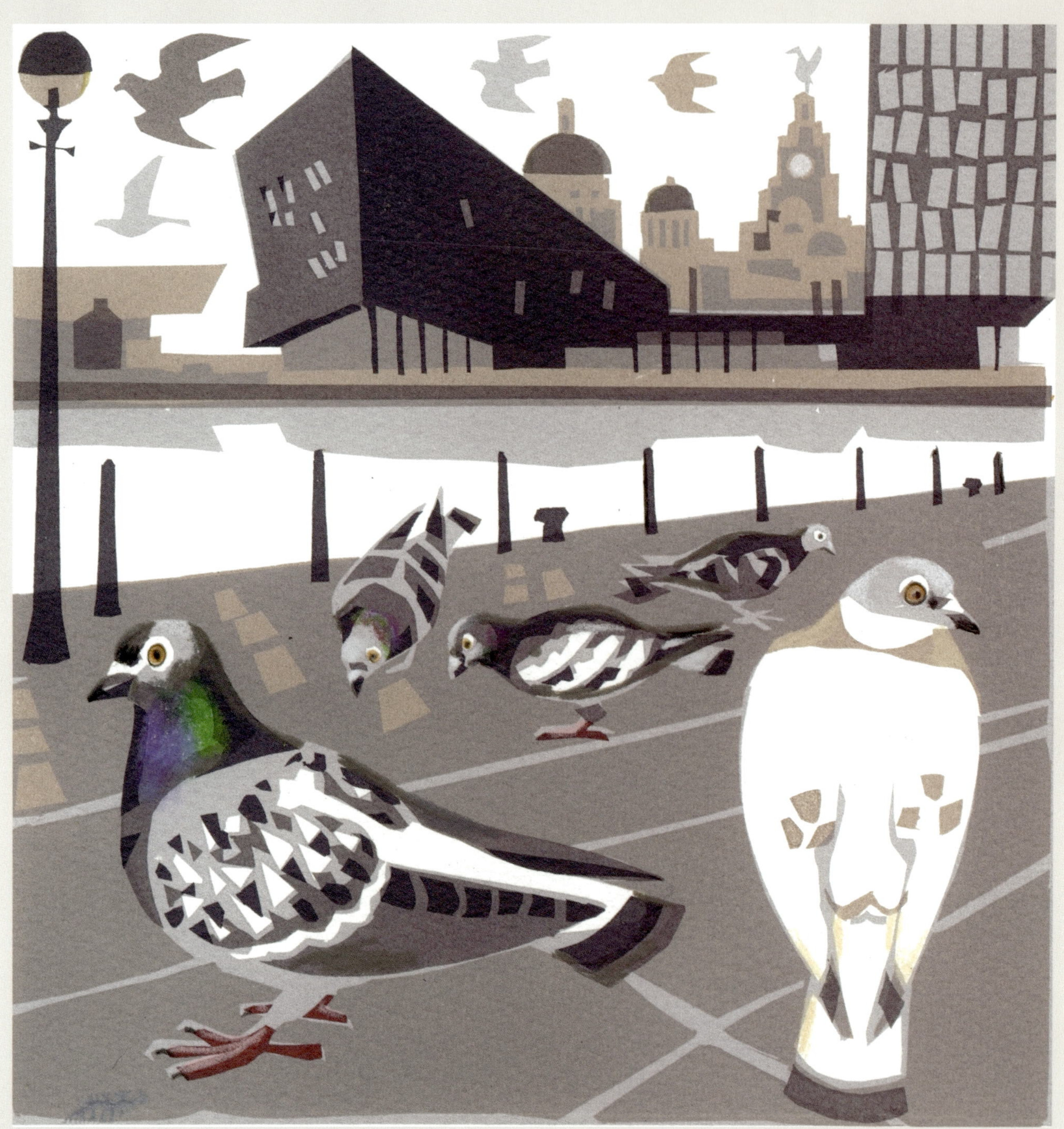

THE FERAL PIGEON

The 228 pigeon hybrids, including tumbler, pouter, fantail, carrier, racing and the feral (*ferox*, wild) farmyard or city (Carry Akroyd's a Liverpudlian) pigeon, are all classified *Columbia livia* – each originating in the sea-cave dwelling rock dove, the purest-bred UK colonies now most abundant in north-west Scotland and Ireland.

Pigeons are the oldest domesticated bird. They provide food, sport and carry messages and pollution monitors. 'Pigeon-hole' derives from dovecotes and pigeon-houses, which still deliver squabs (nestlings) to market. Knowing the definition of 'squab' helped Judith Keppel become the first winner of ITV's *Who Wants to be a Millionaire*.

Until the 'clay pigeon' they were prime shooting targets; and their homing instinct has been refined to produce racing champions; a regal sport for aeons. Elizabeth II has won every major UK pigeon race, her 160 birds newly housed in a £40,000 loft at Sandringham. In March 2019 the pigeon Armando fetched a new world record price of $1.40m. No wonder the UK's 1,000 peregrine pairs so anger the UK's 42,000 pigeon owners, pigeons the peregrine's favourite prey.

The bird's homing ability has made it emblematic of hope, as Noah's dove returning to the ark with an olive leaf attests (Genesis: 8:11). The Holy Ghost is a white dove, universal symbol of peace and love. Picasso's father was a pigeon painter, once a popular genre, and Picasso's drawings of doves of peace for UNICEF *et al.* are even more iconic than those of Georges Braque, his fellow founder of Cubism. He even called his daughter Paloma (Dove).

'Bird brain' is a well-proven misnomer. There are still 'pigeon posts' – Reuters began as one – and in Afghanistan, carriers are banned by the Taliban. Their military importance in both world wars is commemorated by Bletchley Park's 'Pigeons at War' display. They can travel by Underground and research at Canada's Western University proves they can grasp probabilities through number recognition – the first non-primate to do so. Criminally, pigeon corpses stuffed with drugs, etc., are lobbed by accomplices over prison walls.

All these strands meet in the derided feral: Tom Lehrer wrote the song 'Poisoning Pigeons in the Park', and Woody Allen's script for *Stardust Memories* branded them 'rats with wings'. Recent years have seen them outlawed in the world's cities, but they have notable champions. Marilyn Monroe paid urchins to release the ones they had caught for the New York meat market. Henry 'Blowers' Blofeld made them a much-loved feature of his cricket commentaries.

Ferals can be radiantly plumaged and are wonderful flyers. They clean the streets of our takeaway scraps and are clever. For many of us, they are our closest bird – their crooning is a comfort, their antics amusing.

Every day I see from my window
pigeons, up on my roof ledge — the males
are wobbling gyroscopes of lust.

Last week a stranger joined them, a snowwhite
pouting fantail,
Mae West in the Women's Guild.
What becks, what croo-croos, what
demented pirouetting, what a lack
of moustaches to stroke.

The females – no need to be one of them
to know
exactly what they were thinking — pretended
she wasn't there
and went dowdily on with whatever
pigeons do when they're knitting.
NORMAN MacCAIG, 'Wild oats'

Yet, for all their numbers, you rarely see a dead one. They seek the dark to die, true to their cave-dwelling heredity.

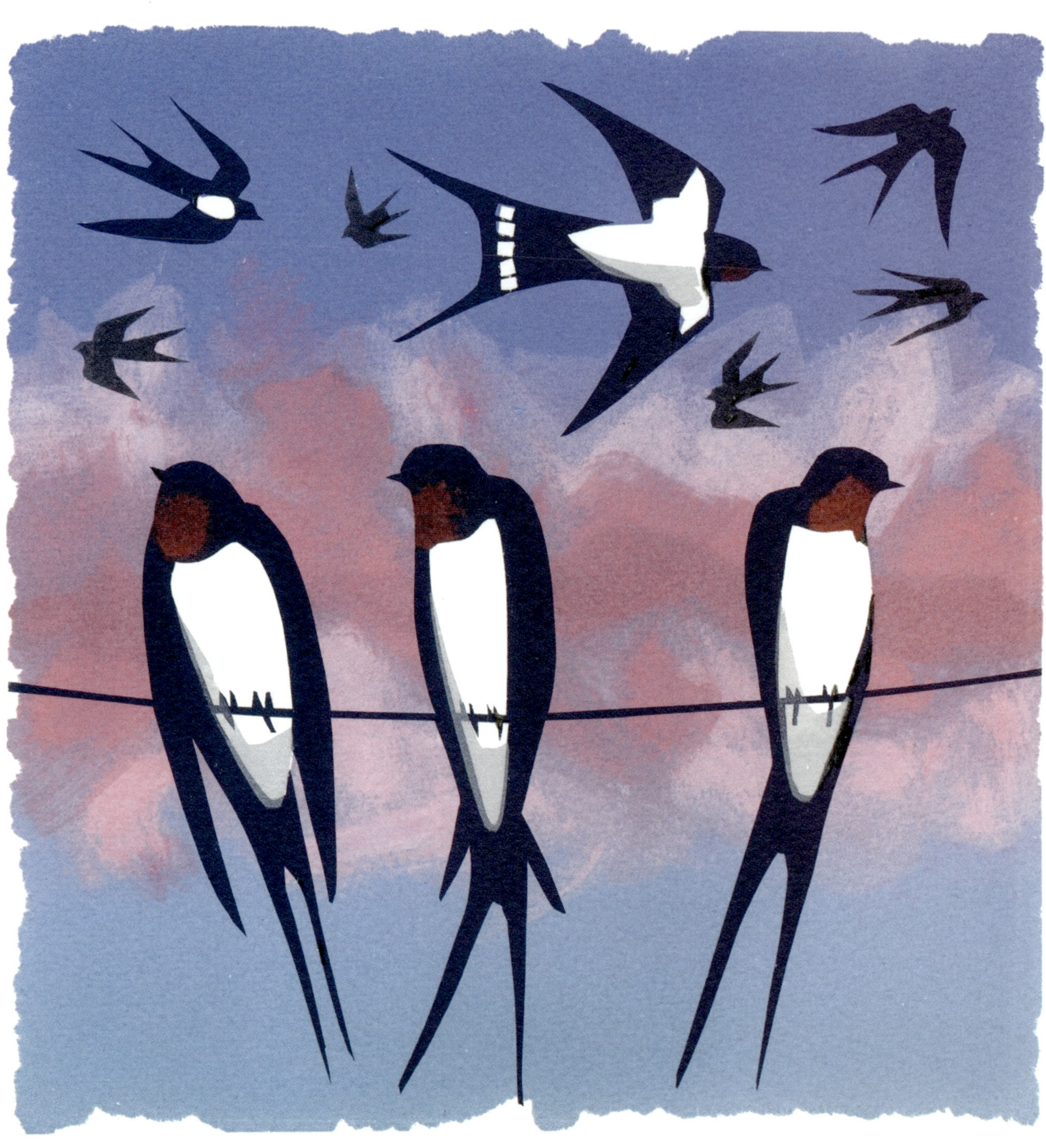

THE SWALLOW

HEW AINSLIE, from 'It's dowie at the hint o hairst'

I remember a blustery autumn day with a succession of single migrating swallows flying south at ground level to avoid the buffeting headwind. How could such a small bird, an ounce (28g) in weight, reach even France after this strength-sapping start to its epic journey? But apparently what swallows hate is a following wind greater than their 37.75 mph calm-air average speed.

It is on still, insect-friendly autumn days that one can sometimes enjoy the spectacle of the migrating frenzy of the swallow (*Hirundo rustica*) and martins, an excitement the Germans have a word for but we do not, *zugunruhe*. A reminder of how spectacularly audible and visible swallows can be at the back end of the year, when they congregate for the migration to southern Africa; some travelling to the very tip of the Cape of Good Hope, one of the longest journeys of any songbird.

The best chance of seeing a British *zugunruhe* is at their autumnal gathering points in Ireland opposite the Welsh peninsulas or in England closest to France. Swallows are our bluest bird after the kingfisher, so Vera Lynn was right: 'blue birds' can be seen 'over the white cliffs of Dover'. *Zugunruhes* do not last long. I arrived during one at the house of the artist Annie Soudain. Overlooking the Channel, it is indicatively near where the pioneering aviator Amy Johnson once took off. Suddenly the excited chittering ceased; summer was over for another year.

Ireland, where grazing cattle still abound, is now the favourite haunt of British swallows. To feed its young a swallow has to fly up to 400 miles (644km) a day at speed. To do this successfully it 'requires', in the words of my *Oldie* magazine colleague James Le Fanu, 'the highest manoeuvrability "index" of any bird…Those elegantly elongated outer tail feathers – or "streamers" – by which the swallow is so instantly recognisable, provide its crucial advantage' (*The Oldie*, February 2019). Swallows are showing a tendency to leave later, a few even seen to winter here; not hibernating underwater as our forefathers thought (see *Dr Johnson's Dictionary*). Insect-friendly mild weather can support a wintering swallow.

Swallows were cave birds before we provided barns, eaves and telegraph wires. Some still have this atavistic tendency, as witness a nest 56 feet (17m) down the shaft of an abandoned Cornish tin mine. But from barns and telegraph wires they are now inseparable.

It is a companionable bird. In Spain, where swallows are welcomed to nest inside houses, it is thought to bring good luck. I have sat in an Andalusian sitting room with half a dozen nests, the birds flitting in and out. British weather does not favour permanently open windows, and perhaps we are also more squeamish. Geoffrey and Judy Grimes let a pair raise two broods in their Kent bedroom and kept a diary. The last entry is 18 September: 'No bedroom visitors for the first time. Heaved a sigh of relief – spring-cleaned the bedroom and closed the window.' The diary begins on 14 May! (*Birds Britannica*).

JOHN H. F. McEWEN, 'Migrants'

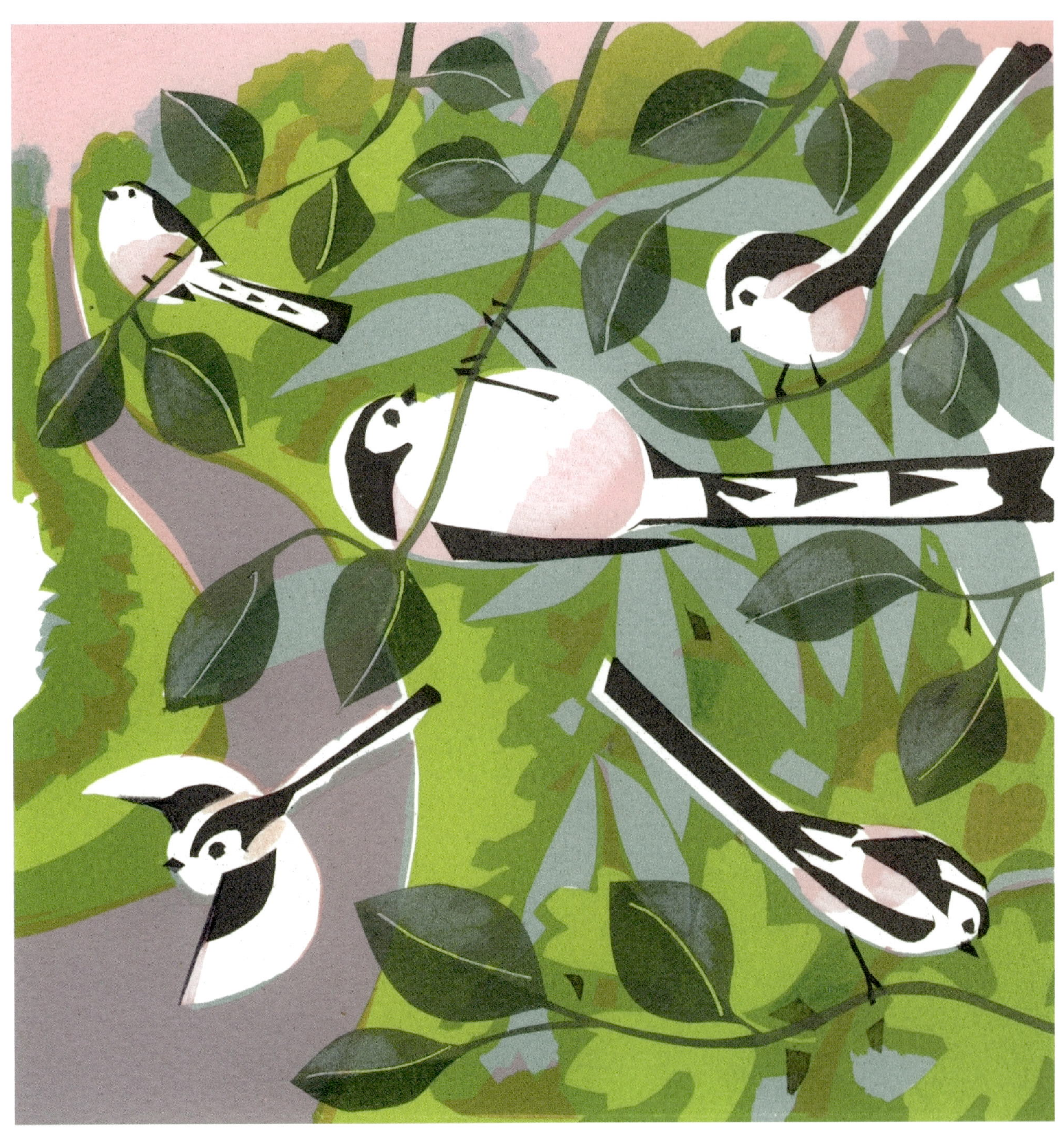

THE LONG-TAILED TIT

'Tit' is short for 'titmouse'. One of the first signs of autumn is the visit of a family of long-tailed titmice, colloquially 'bumbarrels', to the birdbath. In they swoop to create a sudden fountain and, once soaked, return to a convenient perch for a drying flutter before repeating the performance a dozen times. Then, as suddenly, they move on, acrobatically upending as they make their busy way from one tree's canopy or hedge to the next, in their ceaseless wandering. Rarely a ground feeder, it is an agri-phobic bird, never crossing a field but progressing in open country via woods and boundaries. The only places it avoids in Britain are the mountains and treeless islands of the north-west Highlands.

The long-tailed (*Aegithalos caudatus*) is the most exotic of the titmouse family; the similarly long-tailed bearded reedling (formerly bearded tit) is no relation. Long-taileds are always exciting to see, not least for their piebald, flushed-pink, plumage. The 3-inch (7.6-cm) tail disguises that it is one of the tiniest birds – the weight of a five-pence piece. They are ubiquitous, as happy to rove in town as the country, and have increased in Britain by 80 per cent (most of it in urban areas) over the last 25 years. The current population of 330,000 places it fourth of Britain's seven titmice. Lack of fat and an insect diet make it particularly vulnerable to long, harsh winters: 1962/63 was especially devastating, which may explain the population surge and tendency to nest – now commonplace in February – on average three weeks earlier than 50 years ago. It often raises two broods, each usually up to 12 in number, although 20 have been recorded.

The nest is another exceptional feature, a domed marvel of avian architecture built by both birds. John Clare devoted a poem to it:

The oddling bush, close sheltered hedge new-plashed,
Of which spring's early liking makes a guest
First with a shade of green though winter-dashed —

There, full as soon, bumbarrels make a nest
Of mosses grey with cobwebs closely tied
And warm and rich as feather-bed within,
With little hole on its contrary side
That pathway peepers may no knowledge win
Of what her little oval nest contains —
Ten eggs and often twelve with dusts of red
Soft frittered —

From 'Bumbarrel's Nest'

The oval nest is the size of a large orange, camouflaged with a crust of ashen lichen. It is bound to, not supported by, a branch or trunk. Half the cushioning insulation of feathers are from the woodpigeon. Feathers are often plucked from bird carcasses, regardless of species. In one case the feather count included three from mallard. It would be nice to think such discrimination was dictated by taste. 'The outside is as functional as it is beautiful – "camouflaged" and "tiled" with up to 3,000 flakes of lichen, that stick like Velcro, effectively waterproofing it' (*Nature's Home* magazine, spring 2019). The nest, which can take as little as three days to build, has long been deconstructed in the cause of science. William MacGillivray, the nineteenth-century ornithologist, counted 2,379 feathers. Henry Douglas-Home in 1977 cited 3,383 as the record. Unlike shorter-tailed birds, the hen sits facing inward leaving its tail sticking out. Broods can burst the nest, leaving the hen exposed on what looks like a chaffinch nest.

— and full soon the little lanes
Screen the young crowd and hear the twitt'ring song
Of the old birds who call them to be fed
While down the hedge they hang and hide along.

From 'Bumbarrel's Nest'

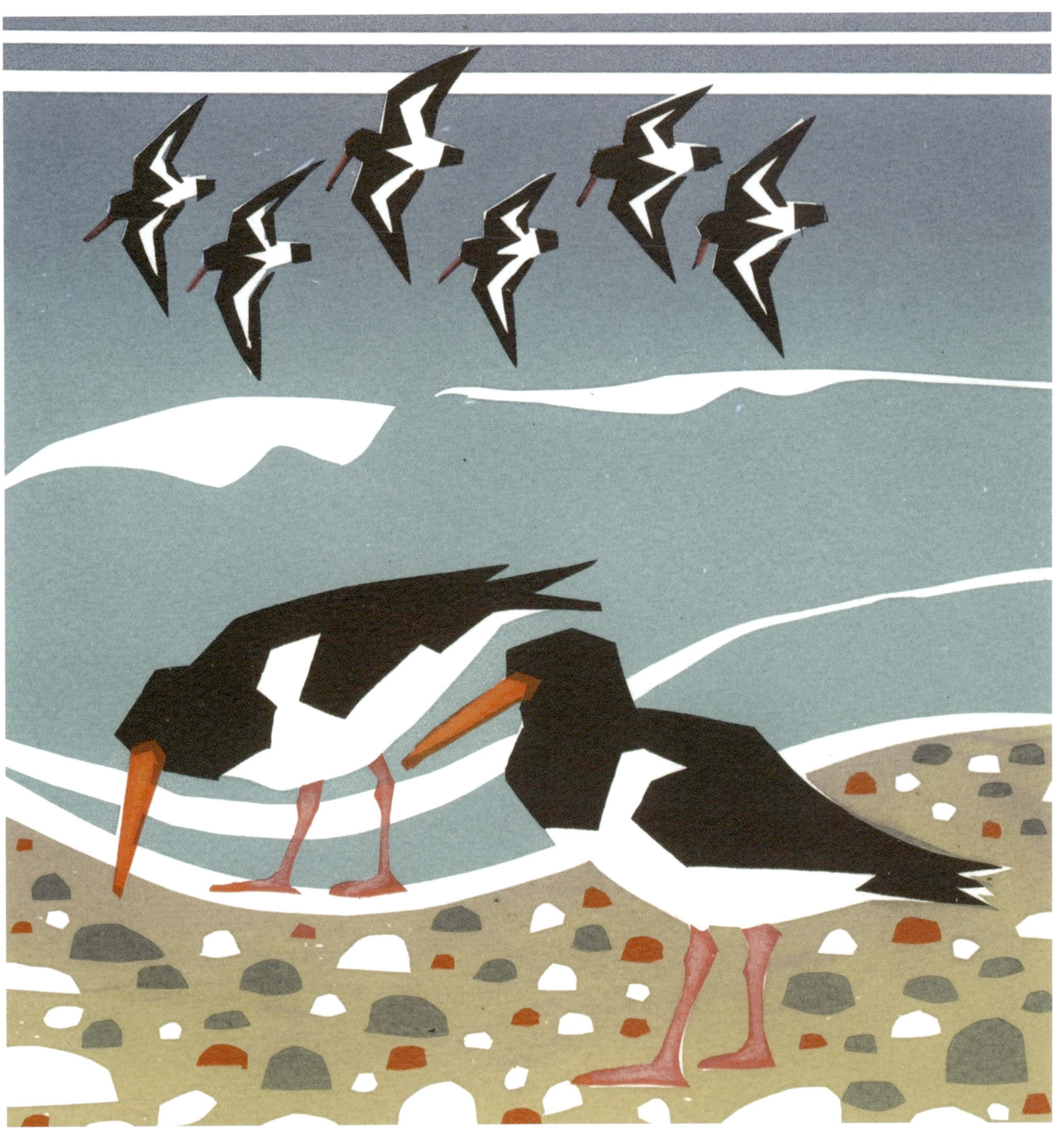

THE OYSTERCATCHER

Oysters are not the preference of the oystercatcher (*Haematopas ostralegus*), but its carrot bill makes short work of almost anything else edible on the foreshore. Colin Tudge (*The Secret Life of Birds*) likens this formidable implement to a Swiss army knife: 'It's a probe for prising molluscs out of the mud. It's a crowbar, for levering the shells of bivalve molluscs apart – or for flicking limpets off their rocks. It's a pair of kitchen scissors, for slicing through the adductor muscle that keeps the bivalve shells together. It's a hammer, for peremptorily smashing the shells of molluscs that refuse to cooperate.'

An oystercatcher can eat a cockle every 72 seconds and a daily weight of nearly one pound (454g) of meat (more than 500 shells).

Its old name is more accurate: the seapie, as in magpie; the dramatic black and white plumage offset by the carrot bill and pink legs. Thomas Southwell, in a 1902 footnote to the seventeenth-century Sir Thomas Browne's book *The Natural History of Norfolk*, wrote: 'it breeds occasionally about Wells, where it is universally known as the "Dickey-bird".'

Like many of our resident birds, the population is significantly increased by winter migrants from mainland Europe; with the oystercatcher, it is doubled to more than 300,000. These roam in flocks, but for most of us, oystercatchers are a seaside-holiday pleasure.

On the nine-hole links at Girvan in Ayrshire, my flamboyant partner hooked his drive out to sea, where the ball hit an almost submerged rock and rebounded onto the fairway. Two oystercatchers were unmoved by this startling event as they stood in twin tidal pools – 'they are stupidly trusting birds', 'BB' (*Tide's Ending*). A breath of wind ruffled one reflection into bright abstraction, while the other remained glassy true to the bold original. Summer days.

In Britain it breeds inland on river shores, the more unfrequented – and therefore usually northern – the better, its eggs blending with the shingle.

…On the Dee's shingle, one swooped at me like a skua; and the piping! — shrillness lancing the brain cells. Yet you're handsome too, black and white as your nature. And that iris — pure ruby. Quick-stepping along in your early pairs in the spring on the beaches, heads down, beaks splayed wide, kleeping like crazy! The eggs like big pebbles in a scrape on the shingle, or in fresh-seeded barley or turf. One I knew, in the hollowed top of a big fence post, beside the Candy Burn. Three eggs on wood-chippings; the bird brooding on its fence post high-rise. Imagine the chicks, their peering, incredulous puff-dried, over the rim, and saying Hey, Maw, what *is this? An early crash course in the art of the parachute, without one. Then the leap of these three fluff-balls, so light it's their only defence, into space.*

GERRY CAMBRIDGE, from 'Haematopus ostralegus'

In Aberdeen, the synthetic beaches offered by the gravel-covered roofs of the Royal Infirmary were soon colonised by the birds to the extent, including other similar roofs, of 250 pairs. If buildings offer this amenity they will be colonised; as is the case with other newly exploited habitats – Midland gravel pits, even Gerry Cambridge's rotten fence post.

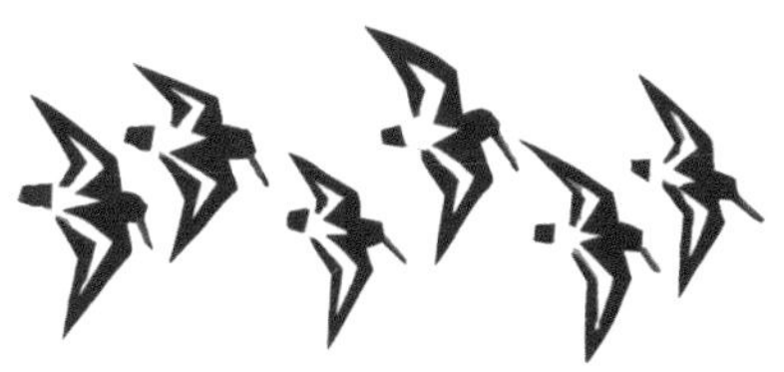

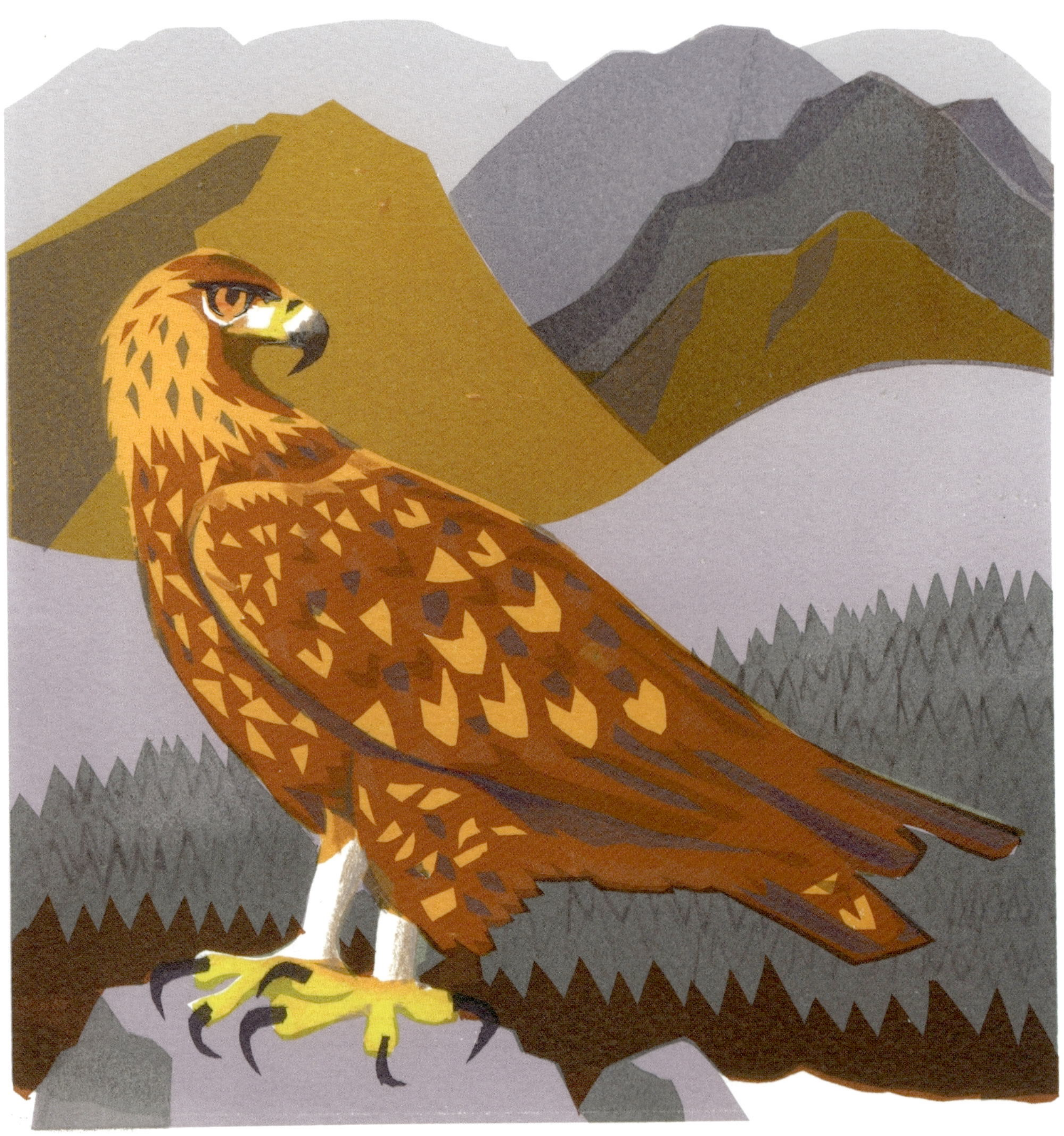

THE GOLDEN EAGLE

He clasps the crag with crooked hands;
Close to the sun in lonely lands,
Ring'd with the azure world, he stands.

The wrinkled sea beneath him crawls;
He watches from his mountain walls,
And like a thunderbolt he falls.

ALFRED, LORD TENNYSON, 'The Eagle'

Portia Simpson, Scotland's first ever female gamekeeper, having gralloched and dragged a stag on the Isle of Rum, took a break and dozed off. There was a 'rush of cooler air'. Opening her eyes, she saw a golden eagle (*Aquila chrysaetos*) 'its talons outstretched towards my torso.' She screamed and jerked up her arms. The eagle veered aside. 'It was the first and last time that I ever fell asleep out on the heather.' (*The Gamekeeper*)

On Islay, the golden eagles kill feral goats by chasing them over cliffs, but the bird's usual diet is rabbits, mountain hares, birds and especially energy-saving red-deer carcasses; hence the mistaken attraction of the sleeping Portia Simpson. *Winterwatch* showed their powers of ferocious consumption: a 66lb (30kg) red-deer carcass stripped bare by two eagles in 7½ hours over several days.

It is all in the Book of Job: 39:28-30

She dwelleth and abideth on the rock, upon the crag of
the rock, and the strong place. From thence she seeketh
the prey, and her eyes behold afar off. Her young ones
also suck up blood: and where the slain are, there is she.

Winterwatch also showed a golden eagle defending a carcass from a fox. The eagle won. No wonder we recognise it – the most globally widespread of all eagles – as the all-conquering king of birds: that there have been eagle-gods far into history; that it was sacred to Zeus, god of the heavens, and his Roman counterpart Jove.

The eagle is the symbol of the fourth evangelist, John, whose gospel opens with that resounding declaration: 'In the beginning was the Word.' Hence the Bible-bearing church-lectern in the form of a brass eagle standing on a globe. The Word will be borne to the four ends of the earth, bird and message inspiring in their majesty. The eagle is among the four 'living creatures' around God's throne (The Apocalypse, 4:7); the symbol of Christ's Ascension into heaven.

The British golden eagle population is growing, but its 508 pairs (2015; 442 in 2003) are mostly confined to north-west Scotland and the Outer Hebrides; the last English bird died in the Lake District in 2016. Efforts are being made to increase its sparse numbers in the eastern Highlands and token representation in Dumfries and Galloway. Plans are also afoot to re-introduce it to Snowdonia.

'Eagle-eyed' we rightly say, its human-sized eye having 340-degree vision and five times a human eye's power. It has proved too strong and fierce for conventional falconry, the 200 mph stoop surpassed in velocity only by the peregrine. Martin Whitley's Dartmoor Hawking offers hunting from horseback with his golden eagle Artemis – the one place outside Mongolia where this sport happens.

The RSPB, SNP and other conservation-minded groups seek to ban game shooting, especially in the interest of raptors, eagles not least. Portia Simpson must have the last word: 'If they didn't use the land for shooting, they'd use it for farming or forestry, which would be devastating for wildlife.'

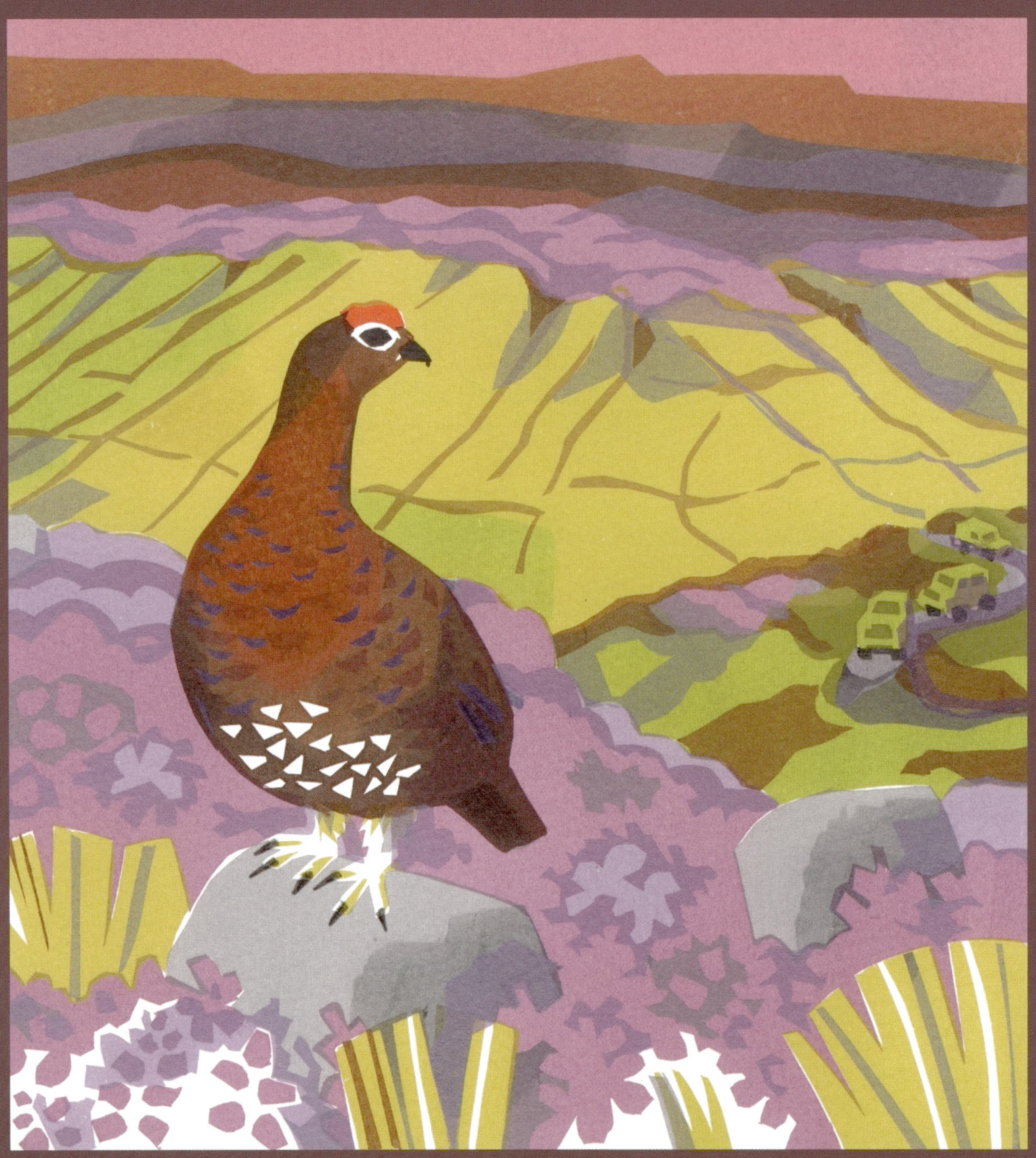

THE RED GROUSE

The red grouse (*Lagopus scotica*), so called after its feather-furred legs, is a subspecies unique to Britain. Gloag's 'Famous Grouse' whisky, advertised by Rodger McPhail's emblematic cock bird, symbolises romantic Scotland worldwide.

Renowned as 'king of game birds' – hardest to shoot, best to eat – restaurants vie to get one to table the day the season opens on 12 August. For pink perfection, philanthropic chef Joe Hardwicke recommends a pre-heated very hot oven and 20 minutes cooking (250° C/ 482° F/ gas mark 9) wrapped in unsmoked bacon.

The kingly title suits its defiance. A cock made life such hell for the bicycling Tomintoul postman he was given a car (*The Birdman*). The cock's staccato 'Go back! Go back!' says it all. Grouse are too wild to rear, unlike pheasants and hybrid red-legged partridges. A hen, more Glenlivet or Glenmorangie in hue than Famous Grouse, brought my car to a halt on a Lammermuir road, as it shepherded its chicks as authoritatively as a lollipop lady.

For W. H. Hudson, early chairman of the RSPB, its 'perfect harmony' with its wild surroundings made it the most beautiful of birds. He understood its 'fascination' for sportsmen: 'it is not the bird only that draws him' but 'the sense of liberty and savage life that returns to man in the midst of mountain and moorland scenery.' How suitable that Heathcliff presents 'a brace of grouse' to Mr Lockwood in *Wuthering Heights*. Lord James Percy, one of field sports' most active advocates, not least as a writer for *Fieldsports* magazine, chose 'romantic' as one of five words to summarise his enthusiasm. One charm of grouse driving is its long interludes. As Alfred Cochrane (1865–1948) wrote in 'The Last Drive':

> *Moved by the beauty of the scene,*
> *While in my butt I wait*
> *Against my fringe of sods I lean*
> *My gun and meditate.*

In the twenty-first century, such dreaming can be troubled. In Brett Westwood and Stephen Moss's (Carry Akroyd illustrated) book of the BBC's popular Radio 4 programme *Tweet of the Day*, it states: 'The paradox is that without shooting and moorland management, red grouse would be at the very least endangered in Britain, and possibly extinct.' By 2005 animal-transmitted parasites and viruses made this a factual threat. Thanks to private funding, an antidote via grit was discovered and delivered, grouse survived and so did grouse moors, to the crucial benefit of dependent and declining nesting waders and upland songbirds. The most successful moor in Scotland, environmentally and for grouse, employs 10 gamekeepers. Such private investment is vital to the economy of remote rural areas, as even the far-from-enthusiastic Scottish National Party accepts.

Opposition to shooting focusses on the protection of the hen harrier, the grouse's chief predator, an inevitably scarce bird at the end of the food chain but an undoubted rarity – its 545 pairs (2018) confined to Scotland. It is a criminal offence to kill it or destroy its eggs, and if this occurs it is the landowner who is the guilty party under the law of 'vicarious responsibility'.

The grouse's wild state makes it prey to population fluctuations, with spring snow and summer heat equally damaging to it in 2018. Many owners cancelled shooting. There used to be 3,000 grouse moors in Edwardian times, with Wales a bastion outrivalled only by Yorkshire. Wales today has none, and the country's breeding – and wintering birds of all species – are 'declining significantly' (*The State of the Birds in Wales*, 2018). In 2019 there are 450 UK grouse moors and the 'Glorious Twelfth' remains the most famous date of the field sports year.

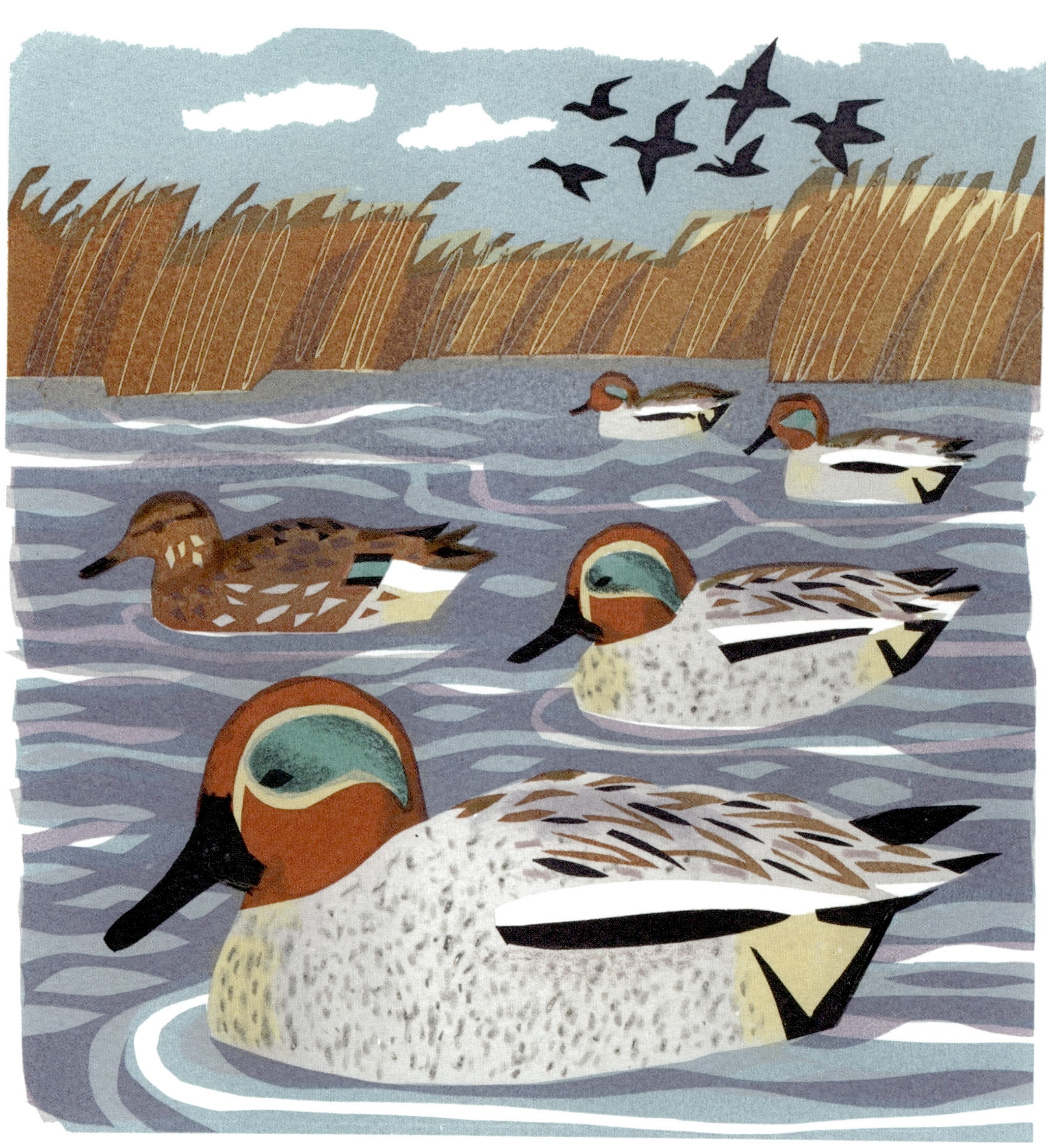

THE TEAL

And near to them ye see the lesser dibbling teal
In bunches with the first that fly from mere to mere,
As they above the rest were lords of earth and air.

MICHAEL DRAYTON, from 'Poly-Olbion'

The teal (*Anas crecca*) has never had any vernacular alternatives to its English name, which imitates one of its calls. It is the smallest of the ducks, the drake as beautiful as any of the jewel-like drakes of the *Anas* family. Visiting Snettisham RSPB reserve, north Norfolk, the green glint of a female's secondary wing feathers beamed like a traffic light from at least 200 yards (183m) Budding couturiers need look no further.

Happily, its numbers are on the increase. October and November witness a winter migration from western Europe and Siberia, which boosts the resident British Isles population of 2,100 breeding pairs to more than 200,000. The resident British population has progressively declined and shifted to the far north and west – Mayo and Galway its Irish strongholds – as population demands (land drainage, pollution) take their toll. By contrast, the winter influx has increased over the last 30 years by 40 per cent in the UK and 30 per cent in Ireland. Choice of country can change for particular birds – perhaps one year England, the next France. Peak passage is in October and November. Outside of nature reserves teal remain wary in the wild. What John Clare wrote in 'To the Snipe' holds true:

Wigeon and teal
And wild duck — restless lot
That from mans dreaded sight will ever steal
To the most dreary spot

Nevertheless, wintering teal can be readily seen on marshes, washes and estuaries. J. A. Baker described them in *The Peregrine*: 'From the big marsh pond came the murmur of the teal flock, like an orchestra tuning up. They were skidding and darting through the water, skating up ripples, braking in a flurry of spray. They sprang into the air as a peregrine came flickering from inland. By the time he reached the pond they were half-way across the estuary, their soft calls like a chime of distant hounds.'

The ability to rocket vertically on take-off explains the collective noun, a 'spring' of teal. *Teal jumping out of the reeds. As ever, the duck leads* is the title of an early oil painting by Peter Scott (*Morning Flight*). Conversely, drakes lead the migration. Teal are monogamous, but change mates annually.

As surface feeders, ice is their enemy; a freeze inland will see them more than ever 'steal' to water protected by rushes or shielding banks. When they spring from such hidden places, even the wiliest hunter may be taken by surprise. Their test as a quarry is as admired as the taste of their meat, for gourmets the rival of grouse or woodcock and, at a third of a mallard, the perfect portion for one.

Yet instinct knows
Not safetys bounds to shun
The firmer ground where skulking fowler goes
With searching dogs and gun

From 'To the Snipe'

Lord Grey found teal the most difficult of ducks to tame. Still, as nature-reserve visitors know, a spring can float and preen obligingly near even to footpaths.

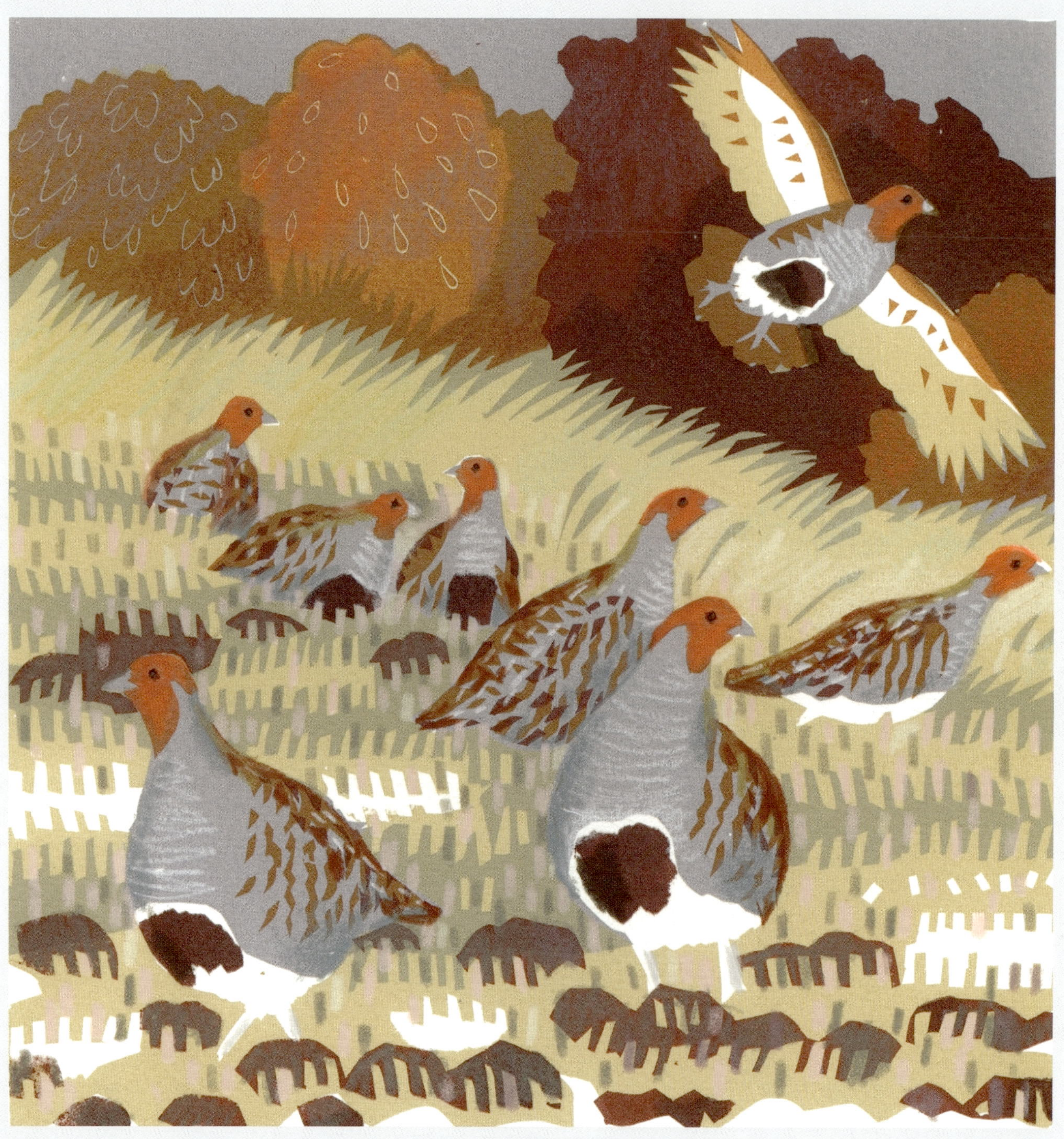

THE GREY PARTRIDGE

Spirits of well–shot woodcock, partridge, snipe
Flutter and bear him up the Norfolk sky:

SIR JOHN BETJEMAN, from 'Death of King George V'.

The grey partridge (*Perdix perdix*) needed no adjective when the 1 September opening of the partridge-shooting season was a sporting date surpassed only by the 'Glorious Twelfth' of August for grouse. 'I must place the partridge second only to the grouse in the sportsman's catalogue of birds which test his all-round skill,' wrote Lord Home, especially when a driven covey exploded 'like shrapnel' (partridge, from Greek *perdesthai*, 'to make explosive noises') over a tree-lined hedge.

What W. H. Hudson wrote in 1895 broadly applied until the 1960s: 'The partridge is a favourite…of all lovers of our wild bird life…he is the only indigenous gallinaceous species in Britain that is not adversely affected by the reclamation of waste lands and spread of cultivation…since he flourishes… where agriculture is most advanced.' Cruel irony: agricultural advance – especially the move from spring to autumn/winter sowing and chemical sprays – has devastated grey partridge numbers. Yet we had been warned. On 7 September 1918 'A. H.' wrote in *Country Life*: 'Partridges have not done so well with us for the past few years…Nowadays farmers keep few objects more constantly before them than the extirpation of weeds and herbivorous insects. Their policy may be commendable from an agricultural point of view, but it means hunger and distress to the little brown birds.'

On a 2015 springtime tour of Jonny Minter's sporting estate in Essex, a winner of a Purdey Award for Game and Conservation, the highlight was a pair of grey partridges. It contrasted with a spring day in 1958 driving through Rothwell, Sir Joseph Nickerson's sporting estate in the Lincolnshire Wolds. Each newly sown field was dotted with grey partridge pairs. In 1952 Nickerson and five guns shot the extant world-record bag, 2,119 (including 50 red-legged partridges). At that date, 2 million grey partridges were shot annually in the UK without diminishing the next year's stock.

'The great day', as Nickerson called it, was achieved on farmland that 20 years before had been treeless and riddled with rabbits and 'vermin'. Even after the war, 80 grey partridges was a good Rothwell bag. This was before Nickerson's successful prescription, detailed in his book *A Shooting Man's Creed*. He employed 14 gamekeepers, planted 250,000 trees and thousands of yards of hedgerow. Kale fields had strips left standing for two years. Feed was scattered from September to April; grit, for digestion, shovelled into hedgerows; 'dusting shelters', for feather cleaning, were placed every 100 yards; nests were discreetly pegged for protection. Visitors to Rothwell, also an award-winning model farm, were rewarded for reporting nests or the presence of vermin.

The contrast with the twenty-first century is inconceivable. In 2019 the grey's entire UK population was 40,000. 'Partridges' sold in shops are Eurasian hybrids, imported, reared and released for shooting, as now happens at Rothwell. Malcolm Brockless, contemporary gamekeeper, has written by contrast of the grey: 'People, whenever…they see the bird closely, comment on its subtle, understated and very "English" beauty'(*Birds Britannica*). Most distinctive is the 'lucky' conker-coloured horseshoe on the chest (bolder in the cock). Llewelyn Powys was another admirer of the bird, especially its fabled fidelity. After a sitting mate was killed by a rat, Powys nightly heard the cock, 'a deluded and distracted lover giving a monotonous utterance to his one simple sentence – "Perdix! Perdix! Perdix! Perdu! Perdu! Perdu!"' (*The Faber Book of Science*), in 2019 the lament for the entire species.

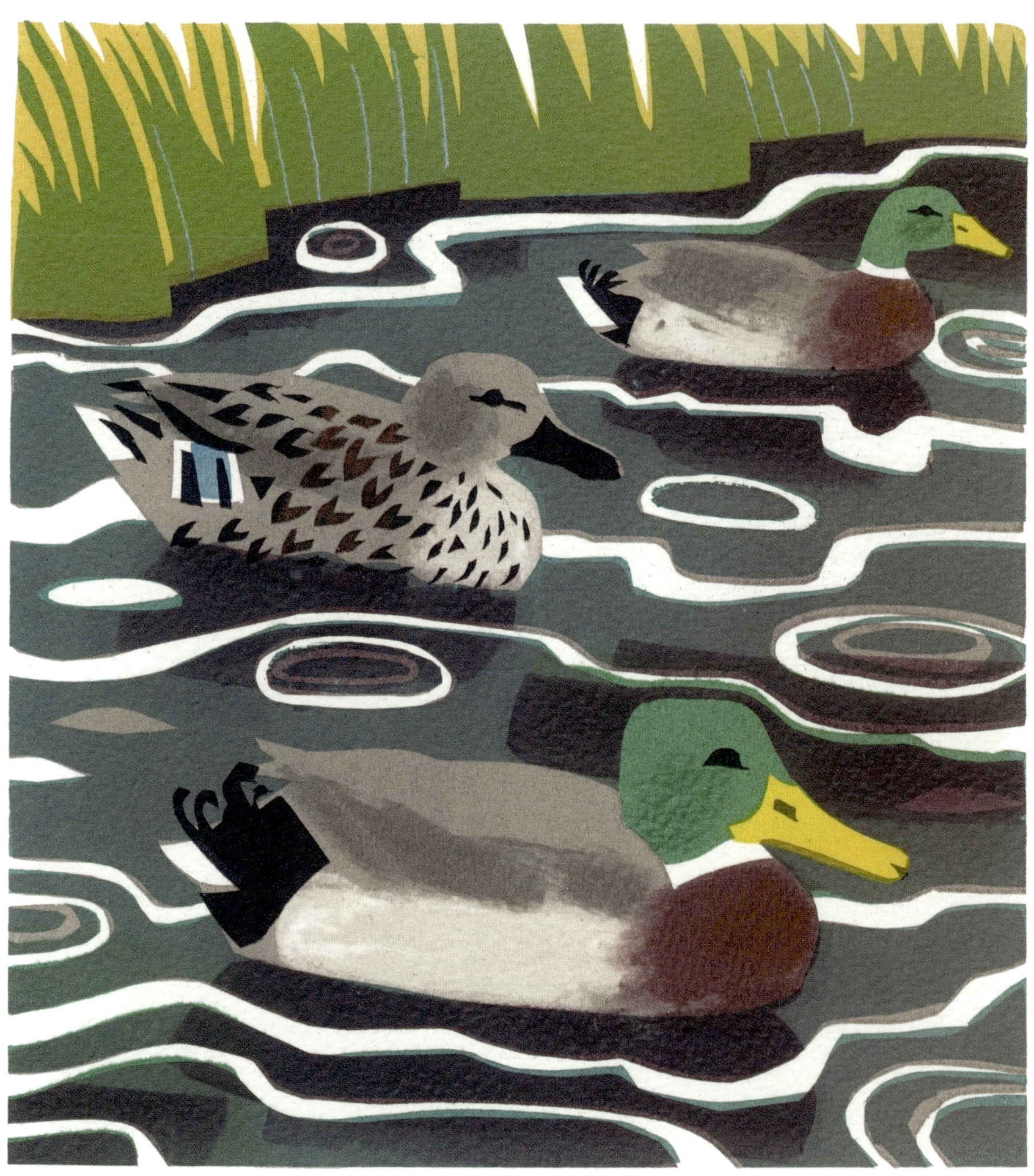

THE MALLARD

When we say 'Let's feed the ducks', we invariably mean the mallard (*Anser platyrhynchos*, flat-billed). When we imitate a duck – 'quack quack' – we mean the mallard; other duck species do not quack, and the duck mallard quacks loudest. The drake has particularly beautiful plumage, with a yellow bill, green refracting blue and purple head, white neck-ring, tawny breast, purple-blue wing-flashes (secondaries) emphasised by parallel rows of white and black, pencilled belly, white cream under-wing coverts, jet-black rump ending in curled 'drake's tails' offset by pointed white tail feathers. Mallard were a favourite bird of Tunnicliffe. He sketched them frequently, and they were the subject of many of his exhibition watercolour pictures. The London and North Eastern Railway's *Mallard*, the fastest ever steam train, took its blue from the drake's wing flash. The record speed – 126 mph – was achieved on Stoke Bank during a journey on 3 July 1938. When it was withdrawn on 25 April 1963, it had covered half a million miles.

Lord Grey had a low opinion of mallard: 'Beautiful as the drakes are in their season, they have a coarse appearance among other kinds of waterfowl. The females especially are common and underbred in appearance and vulgar in manner. They are clever at finding the food that is meant for choicer birds… I endeavour, therefore, to exclude them,' he wrote of his wildfowl collection. 'Bottom feeders!', exclaimed Alexander Hesketh, when I said how exceptionally good mallard are to eat. True, they often upend to feed in water but, as Lord Grey found, they enjoy almost everything, even – it has been recorded, a live sparrow.

Their rough behaviour is also undeniable. Another exemplary Northumbrian countryman, the late High Sherriff Peregrine Fairfax, bred mallard and it was an eye-opener to see the rapacity of the drakes in his pen. Chaucer described 'the drake, stroyere of his owene kind' (*The Parlement of Fowles*). Any city park during the breeding season will show drakes in relentless pursuit of a duck. If the duck is swimming, the weight of a gang of rapacious drakes can drown it. Mrs Beeton considered the murderous drake an urban problem: 'It is to be regretted that domestication has seriously deteriorated the moral character of the duck. In a wild state, he is a faithful husband, desiring but one wife' (*Household Management*).

The mallard has generated 20 officially recognised 'farmyard' breeds since the fourteenth century, among them the Aylesbury and khaki Campbell. Donald Duck too, it must be said. Its omnivorous appetite is matched by its nesting adaptability. Typically a ground nester, as the duck's brown camouflage declares, it is capable of nesting in tree holes and on buildings a long way from water. The duck shepherds the ducklings to relative aquatic safety after they have sometimes dropped from a great height (such as Canada House, Trafalgar Square, 150ft/46m). The female, in common with 5 per cent of birds, has sole parental responsibility. 'In my year at No. 10 Downing Street,' wrote Lord Home of his time as prime minister, 'a mallard duck nested in some low hypericum just outside the Cabinet Room. Eventually she left by the back door, and so a little later did I.'

The mallard is an elusive quarry. Centuries-old domestication has done nothing to diminish the right to its old name 'wild duck'. 'By day he will eat from any person's hand; in the evening he returns to his ancient wary habit', wrote Hudson (*Birds and Green Places*).

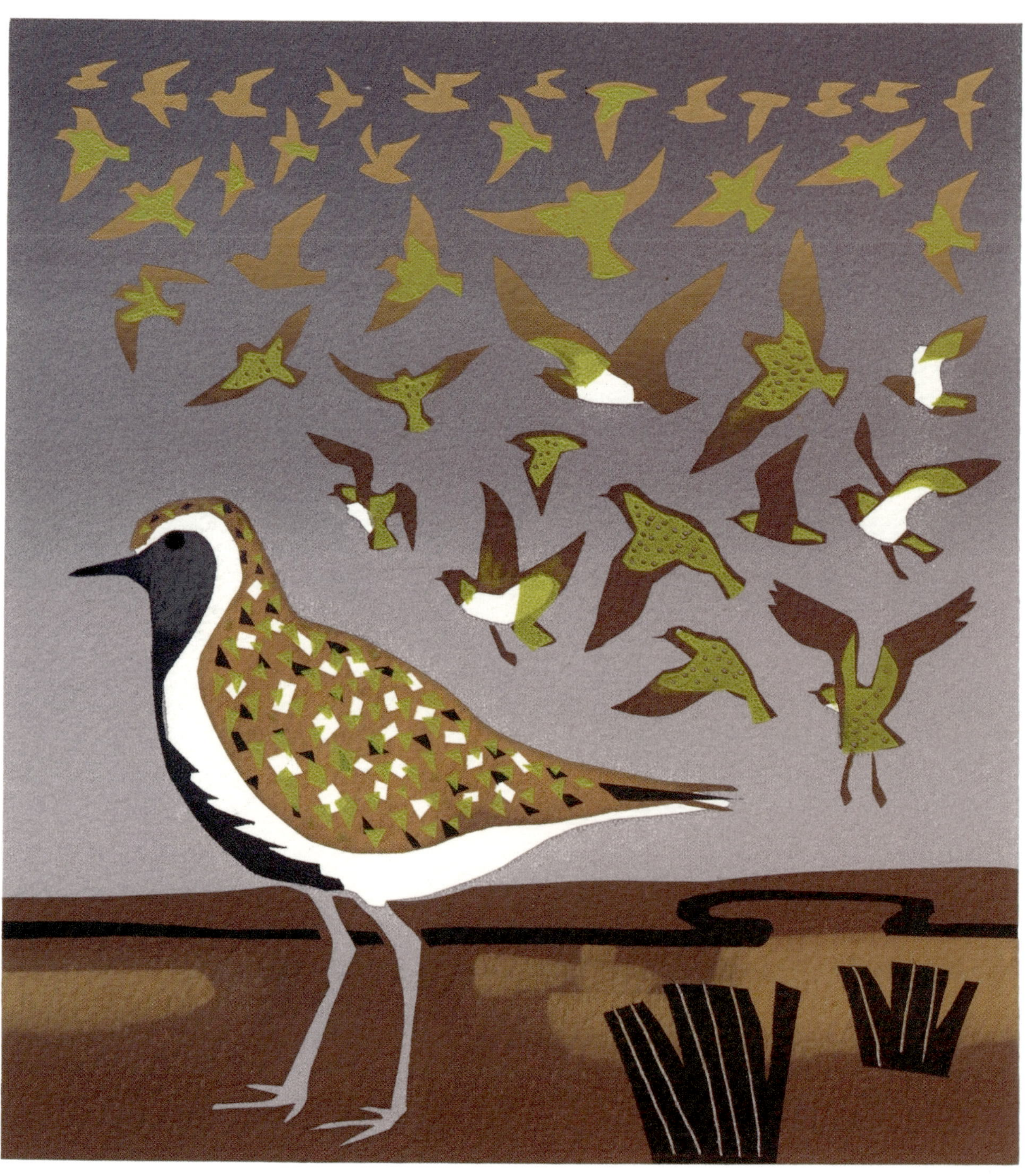

THE GOLDEN PLOVER

Between 1 September and 31 January coots, moorhens and golden plover (*Pluvialis*, rainy, *apricaria*, sunbathe, i.e. sunlit rain) are legal quarry. Who eats coots and moorhens? And golden plover, a famous delicacy, drew a blank at Allens, Mayfair's oldest butcher. The name did not even ring bells. Clearly, the golden is more observed than shot in 2019.

Golden and green plover, so often seen together, do indeed suit golden autumn and green spring. It is when it flies in flocks out of the breeding season that the golden is most spectacular, whereas lapwings are seen at their best when the males perform solo aerobatics over the nesting grounds.

Of course the golden too has much to recommend it in spring. In April 1947 Charles Tunnicliffe and his wife Winifred moved to a shore-side house on Anglesey. They were soon transfixed by the goldens, by then in their gorgeous summer breeding plumage – white winter necks and breasts now jet black, their backs and wings contrasting as white-edged golden shawls. They agreed they were the epitome of elegance. J. A. Baker described a flock feeding in a March corn field: 'listening and stabbing forward and down, like big thrushes…Their black chests shone in the sun below the mustard yellow of their backs, like black shoes half covered with buttercup dust.' The black neck and chest is a deeper hue in the northern birds but not as pronounced as in the grey plover; the golden traditionally described as the grey plover's 'page'. Like the latter and in the even less pronounced case of the dunlin, black is replaced by white in the winter. It makes them blend with other waders on winter mudflats, their breeding colours similarly lost or toned down. The male has a spring song, which 'as he ascends with rapid wheelings high above your head… approaches nearly to the note of a thrush or blackbird', wrote Charles St John (*Wild Sports & Natural History of the Highlands*).

Tunnicliffe reported a September sighting, some retaining their black summer breasts, most in various stages of moult, the freshly plumaged juveniles appearing new-minted among the tarnished adults. All retain the golden shawl, which makes the flash of a flying pack, hence the bird's scientific identity, one of the world's glories. Estuaries from September through to April should offer many chances of seeing it.

My first glimpse was in Baker country, coastal Essex. The September fields were glazed with cobwebs but out to sea rain threatened. Against broken cloud, a pack of several hundred waders sped beyond the tide line. Strung out, they poured across the sky from black on white to white on dark, a sudden twist, glinting bright as a sun-struck windscreen, revealing their golden identity.

Baker found 619 peregrine kills over 10 Essex winters and only two were golden plover, a tribute perhaps to their aerial skills. But St John, 50 years earlier, wrote they were 'a favourite prey' of Perthshire peregrines, affording them 'a severe chase'. He described seeing 'a pursuit of this kind last for nearly ten minutes, the plover turning and doubling like a hare before greyhounds…but in vain.'

Baker wrote: 'As I passed the farm, a flock of golden plover went up like a puff of gunsmoke. The whole flock streamed low, then slowly rose, like a single golden wing.'

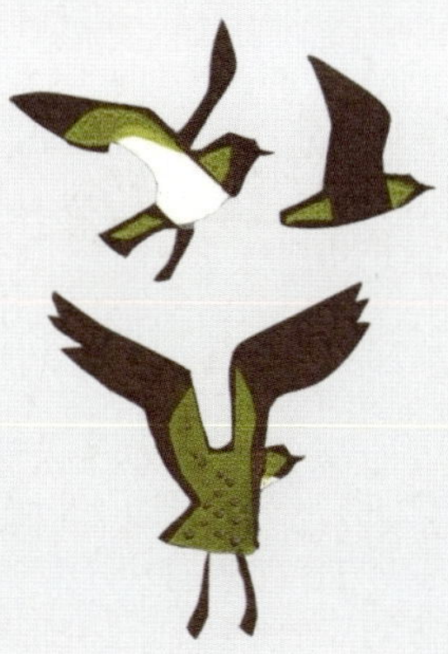

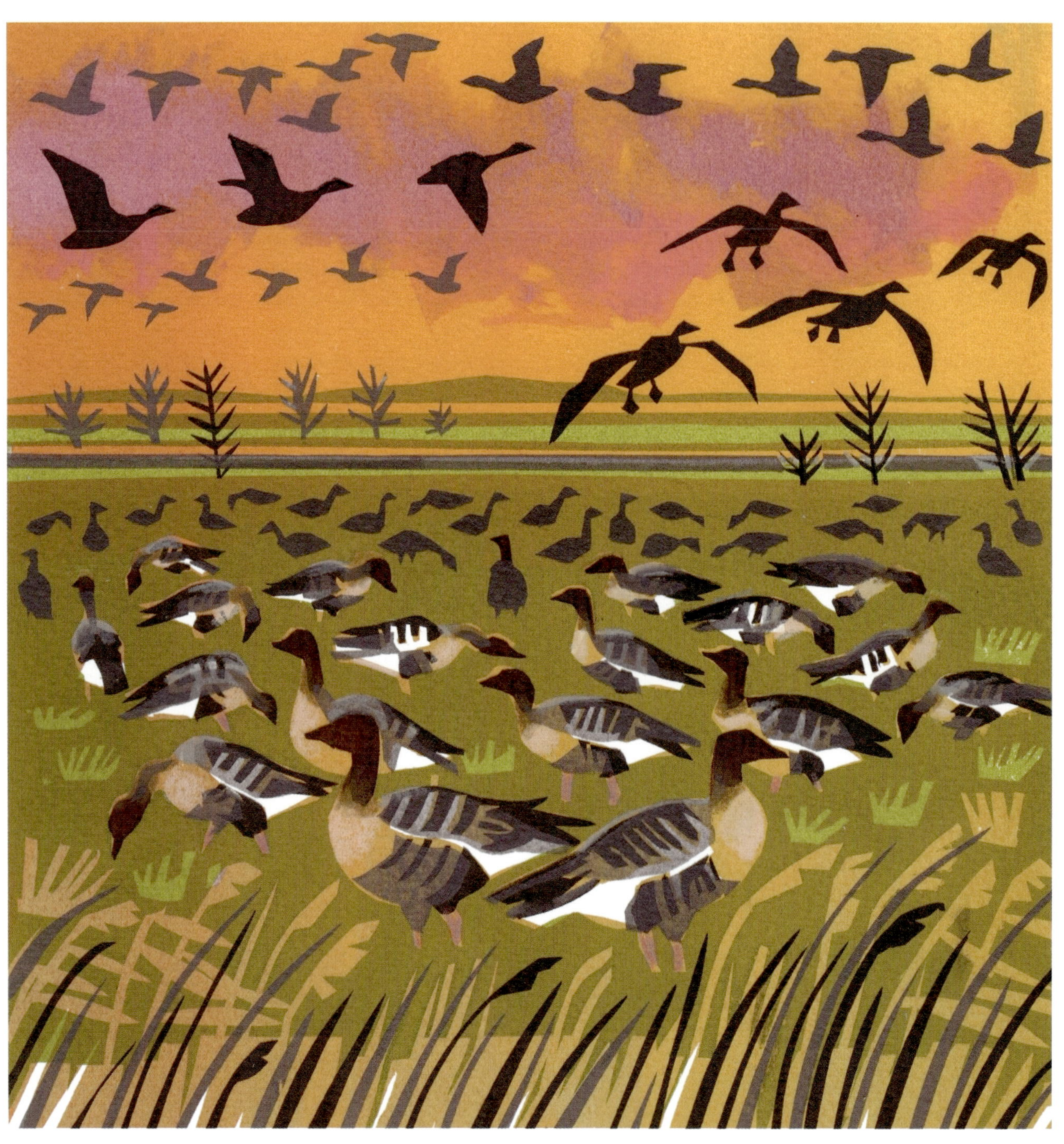

THE PINK-FOOTED GOOSE

ALAN SWARDEN, 'Pinkfeet'

Anyone who hears the clamorous wall of sound created by thousands of airborne pink-footed geese (*Anser brachyhynchus*) will understand Peter Scott's comparing it only to 'the "Sanctus" of Bach's B Minor Mass' (*Morning Flight*). Novelist painter Helena McEwen describes their almost-as-astonishing, whiffling (see The Rook, page 45), descents: 'All at once…there is a hush…I see hundreds of geese…falling out of the sky in the shape of a vast falling leaf' (*The Big House*).

Holkham and other north Norfolk marshes, Martin Mere (Lancashire), and Montrose Basin (Angus) are some favourite locations. No men have done the bird so proud in pictures and words as Scott and his fellow wildfowling artist, naturalist and conservationist Denys Watkins-Pitchford ('BB'). His *Manka, The Sky Gipsy* (1939), a *Moby Dick* tale of Manka, an albino pinkfoot, is a pioneer bird novel. And an albino pinkfoot in Scott's first wildfowl aviary inspired Paul Gallico's best-selling Dunkirk novella, *The Snow Goose* (1941). The pinkfoot is the wildfowler's goose par excellence, the bird which took centre stage for the youthful, wildfowling Peter Scott and 'BB'. The old professional wildfowler Foxy, in obsessed pursuit of Manka, personified the breed: 'Foxy despised what he called 'shore-poppers' – 'these amateur wildfowlers were the curse of his life, and he loathed the sight of them. Every year more of them came, and none of them knew the first thing about shooting. All they did was scare the geese away.'

Henry Douglas-Home told a sinister Second World War wildfowl story. The Home Fleet, anchored intermittently between the Moray Firth and Scapa Flow, was subject to damaging information leaks. The spy was discovered to be a London society dentist, a familiar wildfowling visitor. He had been sending shipping codes to his London control, sewn and hidden in the necks of the geese he had shot.

Pinkfeet breed primarily in central Iceland and arrive in Scotland in early autumn. In the late nineteenth century they began to come inland from salt marshes to feed on farmland – stubbles in winter, pasture in spring – the high protein in new growth strengthening them for the homeward journey. Introduction of sugar beet as a staple British crop, following the 1936 Sugar Industry Act, supplemented their winter diet to dramatic effect. Norfolk is the centre of UK sugar-beet production and in 2019 attracts 100,000 wintering pinkfeet, addicted to the discarded green tops.

In 1946 Scott began catching and ringing wild geese using rocket-propelled nets and founded the Severn Wildfowl Trust, forerunner of the World Wildlife Fund. I witnessed one of these early nettings on a Berwickshire stubble, excitedly galloping across the field to where the captured birds flopped about under the mesh. Their rough legs were ringed, the white and grey-barred tails dunked in a bucket of pink dye. Seventy years on, thanks in large degree to the increased production of Norfolk sugar beet, the British migration has grown from 30,000 to a record 536,871 (2015). After Scott abandoned wildfowling for conservation his art lacked its former poetry – something 'BB' never lost, in pictures or prose: a pinkfoot's neck 'the colour of old snow', an estuary 'shining like a sword' (*Tide's Ending*).

In 1989 I interviewed Sir Peter. I told him 'BB', 83 and on dialysis, had travelled north and, with the help of his friend and unofficial gun-bearer, Badger Walker, had shot his annual Angus pinkfoot. Sir Peter understood.

VIOLET JACOB, from 'The Wild Geese'

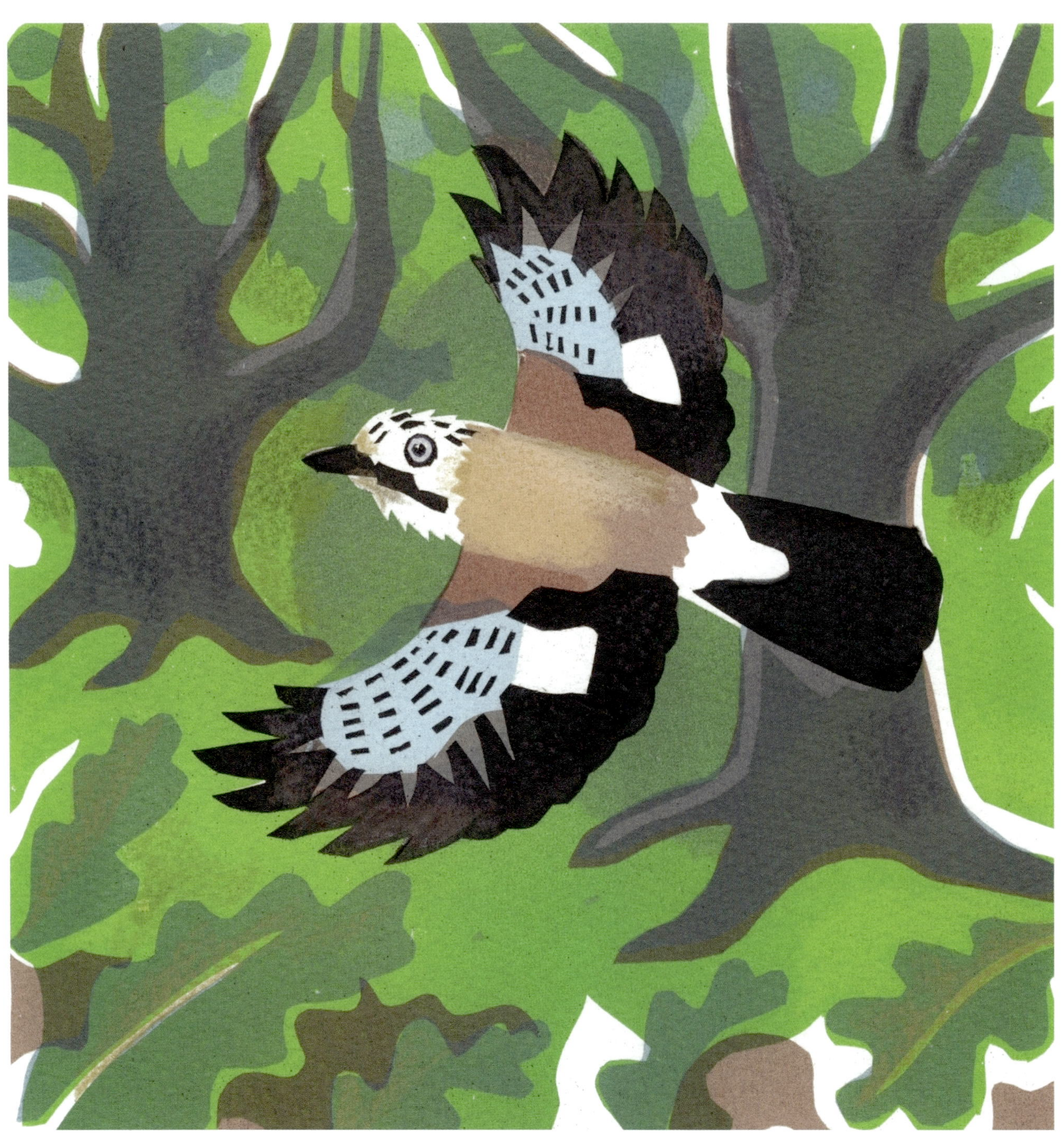

THE JAY

THOMAS GISBORNE, from 'Walks in a Forest'

The jay (*Garrulus glandiarius*, the first refers to its cry, the second to its habit of hoarding the acorn, *glandis*) has been of more economic benefit to Britain than any other bird because where would we have been without the hearts of oak which allowed Britannia to rule the waves, had it not been for the planting of acorns by generations of jays? Five thousand mature oaks went into the building of the Trafalgar veteran 'fighting' *Temeraire* alone.

It has been estimated each jay can store 5,000 acorns, a sublingual pouch enabling it to carry three or four at a time. The acorns are usually buried in damp ground, to be retrieved when food is scarce. Jays have remarkable memories. One bird was seen recovering an acorn after burrowing through over a foot of snow; supplies can last to help feed nestlings the following June. The last official British count (2009) estimated 170,000 pairs; thus 1.7 million acorns are stored annually. Even though most of these are retrieved, that still leaves plenty of potential trees. Jays also bury beechnuts. What is more, burying offers protection since acorns, like many seeds, have no means of calculation to start leaves growing, which is why it is so helpful when jays and squirrels bury beech mast and acorns. Being an inch deep in earth keeps them safe and warm till spring arrives to foster growth (see Peter Wholleben, *The Hidden Life of Trees*).

Their cry is as startling as the beauty of their plumage, alike in both sexes.

BISHOP RICHARD MANT, 'Jay's Scream'

The white rump is as eye-catching as the chequered dazzle of its sky-blue wing coverts, once a coveted fashion item. In 1880 the Duchess of Edinburgh notoriously had a muff made of these small blue feathers.

The jet knib of a beak may be small for one of the crow tribe but is still a weapon. A bird lover once told me that had I arrived a few minutes earlier I would have witnessed a jay tearing apart a mature great tit it had caught. Eggs, sitting songsters, nestlings, chicks are all grist to its mill. W. H. Hudson deplored the decline of jays and 'extirpation' of magpies to the 'stupid and deadly animosity' of gamekeepers. One wonders if he would retain that opinion today, when the decline in keeper numbers and songsters is surely not disconnected from the ubiquity and growing numbers of jays and crows in general. Jays, unknown to London in Hudson's day, now nest in the garden of Buckingham Palace.

They make amusing pets, with a gift for mimicry also used in the wild. Peter Thomsett of Claxton, Norfolk, had a jay that 'could mimic the cat and telephone so well you couldn't tell which was which'. (*Birds Britannica*).

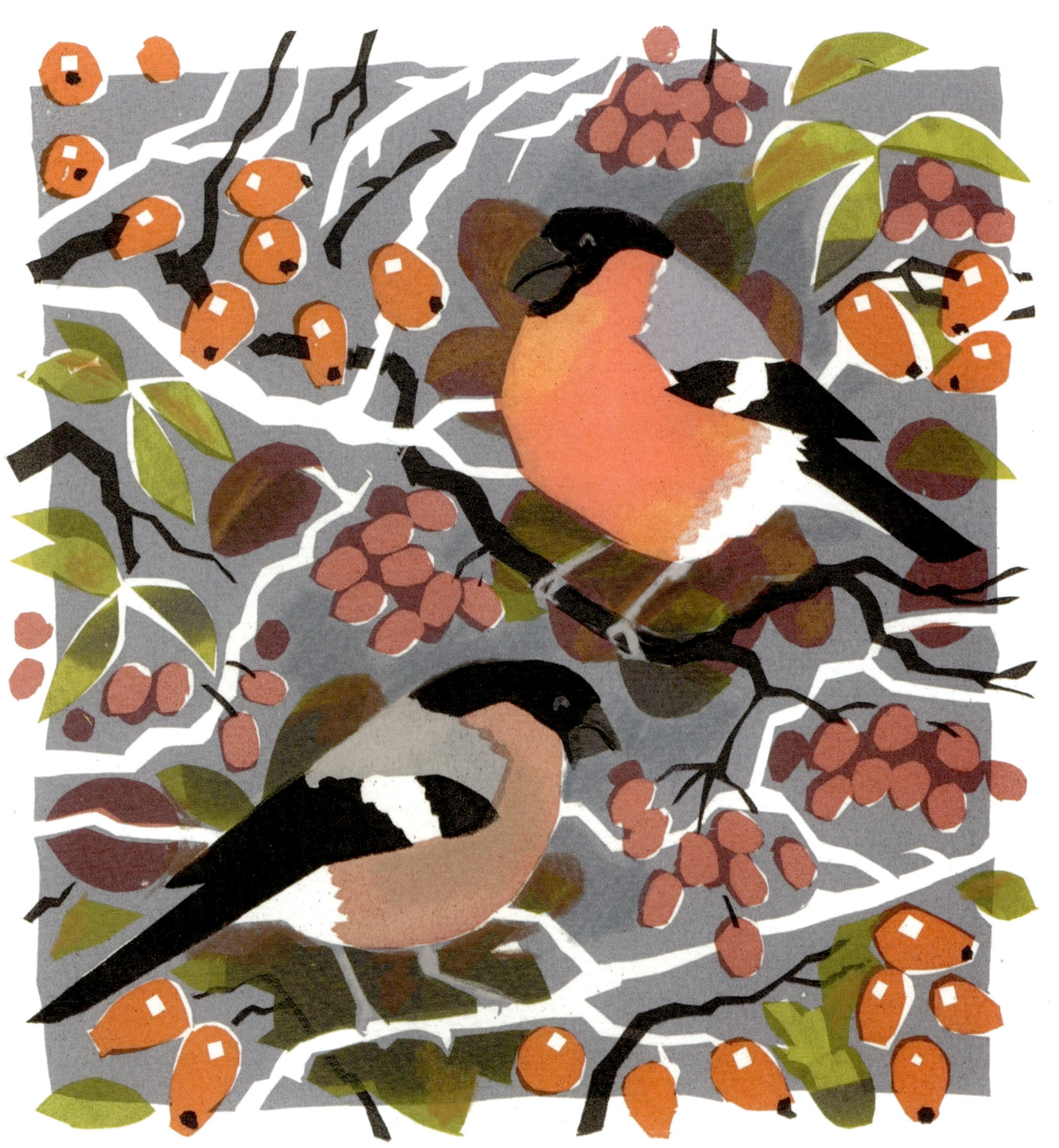

THE BULLFINCH

It was greatly to the advantage of 'BB' as a nature writer that he was brought up in the Northamptonshire countryside and received a long art training. It sharpened his verbal descriptions and lent authority to his aesthetic judgments.

He had a particular affection for the bullfinch (*Pyrrhula pyrrhula*) or 'bully', after its black and stubby bullish beak and head. His favourite pictorial subject was a pair perched on a blackberry branch in a Northamptonshire lane. Carry Akroyd lives in the county and still occasionally sees them in the lanes.

One of his earliest successes was *The Countryman's Bedside Book* first published in 1940 (see also *BB's Birds*). The following passages are typical of those which solaced many an overseas wartime serviceman.

'There can be no prettier sight than a family of bullfinches in late autumn when the red berries glow against the hedge, and the twigs have taken on that soft purple-bloom which seems almost like a mist. The soft pink of the male bird's breast harmonises beautifully with its surroundings…They like deserted roadways and spinneys.'

He found the female's colouring equally pleasing: 'The hen with her soft lavender, black and white plumage is a wonderful foil to the resplendent male.'

The honours of his ebon poll
Were brighter than the sleekest mole;
His bosom of the hue
With which Aurora decks the skies,
When piping winds shall soon arise
To sweep up all the dew.
WILLIAM COWPER, from 'On the Death of Mrs Throckmorton's Bullfinch'

The bullfinch, an able birdsong mimic, was once a popular pet. Rescued 'bullies' were a constant of 'BB''s life. Usually nestlings, he fed them hourly, with the point of a sable brush, on boiled canary seed, passed through a hair sieve to remove the husks, plus ants' eggs and mashed boiled hen's

egg lubricated with a drop of milk. The female broods the young at night until they are fledged, so a warm covering was required.

His favourite was Biddy. 'Her greatest treat was to curl up amongst my hair, when my head was resting on the back of the chair, and in this cosy nest would go fast asleep.' She also gathered 'enormous moustaches of wool and hairs' from the carpet and offered them with a 'trilling twitter'. He recognised this as typical mating behaviour, also a welcoming gesture. Alas, his father accidentally trod on Biddy. Another of the bullies 'BB' tried to rescue, at the point of death sang a melancholy song he had never heard before.

In living memory bullfinches' taste for fruit-tree buds made them a killable pest. They are shy, glimpsed invariably in pairs, but can be tempted to feeders by sunflower and other seeds, particularly in rural gardens. The UK population (220,000 territories) is steady but 40 per cent lower than in the mid-1960s. One recent bird map shows them absent only from the extreme north-west of Scotland, which is where I last saw a pair as we waited, David, Duncan McEwen and I, for the Kylerhea ferry on Skye.

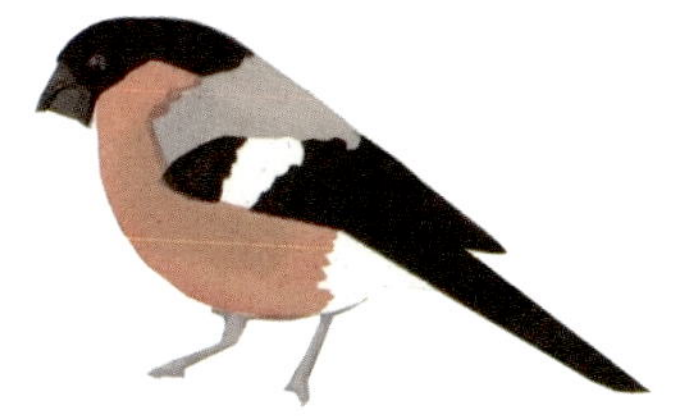

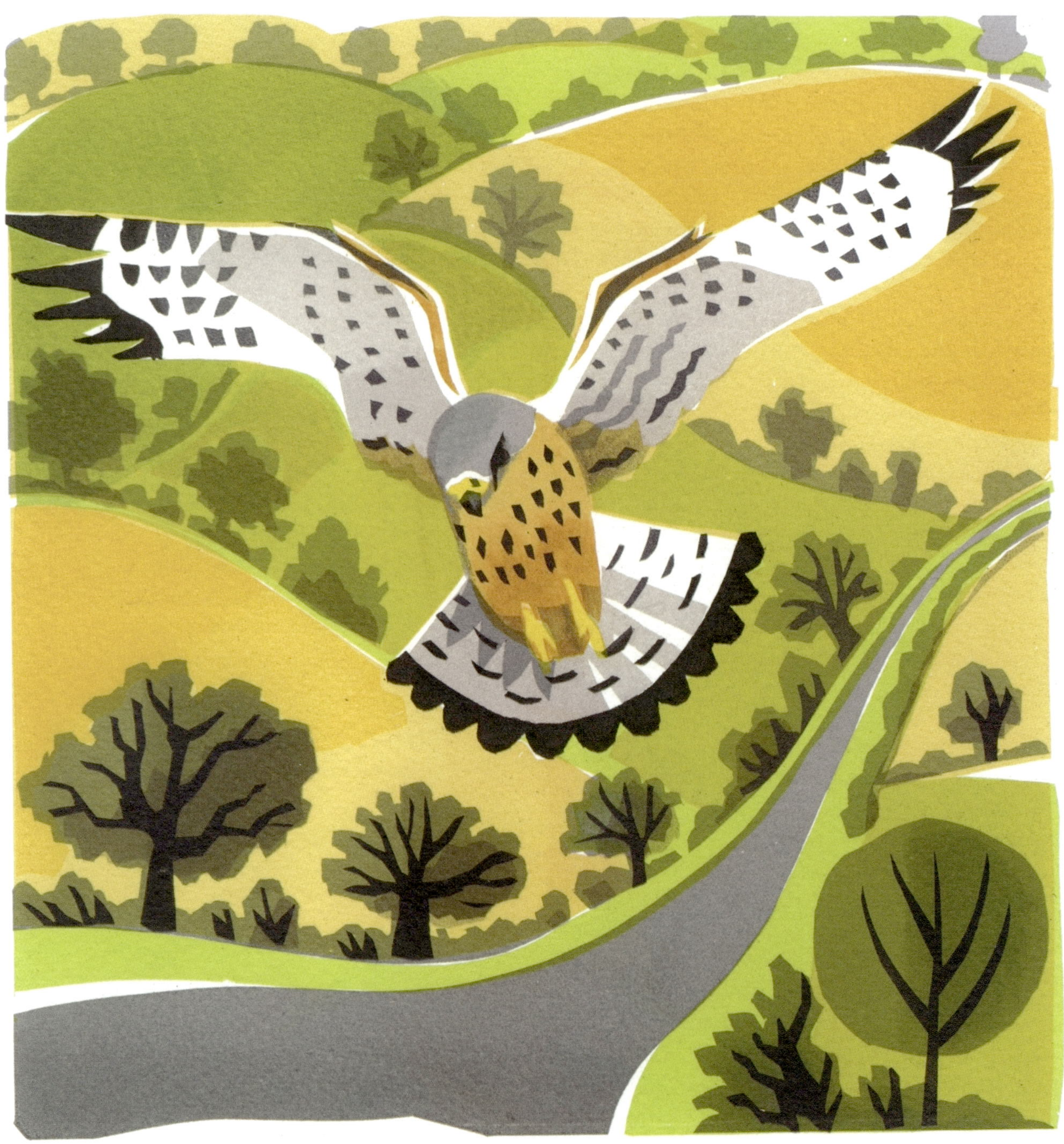

THE KESTREL

I caught this morning morning's minion, king-
  dom of daylight's dauphin, dapple-dawn-drawn falcon, in
    his riding
Of the rolling level underneath him steady air, and striding
High there, how he rung upon the rein of a wimpling wing
In his ecstasy! Then off, off forth on swing,
  As a skate's heel sweeps smooth on a bow-bend: the hurl
    And gliding
Rebuffed the big wind. My heart in hiding
Stirred for a bird, — the achieve of, the mastery of the thing!

GERARD MANLEY HOPKINS, from 'The Windhover'

When this poem was recited in *The Simpsons* Bart went 'Wow!'

Kestrels (colloquially 'windhovers') were thought mere mousers in the medieval hierarchy of falconry, worthy only of a knave. Hence the title of Barry Hines's 1967 novel *A Kestrel for a Knave*, the bird symbolic of freedom for its teenage falconer, Billy Casper, otherwise condemned to a bleak post-industrial future in Barnsley. Successfully filmed by Ken Loach as *Kes* (another colloquial name), it had the damaging effect of encouraging boys to rob kestrel nests. Book and film gave a misleading impression of how easy it is to train the bird. Due to its very low body weight, 8 oz (226g) for a female and 6 oz (170g) for a tiercel (male), finding a kestrel's optimum flying weight is a fine art; even a miscalculation of half an ounce can prove fatal. The tiercel catches the eye with its slate grey, black, white and candy-pink plumage; the female is a speckled brown.

The kestrel (*Falco tinnunculus*) derives its name from the French *crecelle*, 'to rattle'; similarly *tinnunculus*, 'little bell ringer', also refers to its call. The stillness achieved by its hover was tested on BBC Radio's *Living World*. A theodolite revealed the bird could hold its position to within a centimetre for 28 seconds. The ability to see the wavelength of ultraviolet light also allows it to trace the movement of rodents from their urine trails. The proliferation of voles on the banks of new motorways made it the 'motorway hawk' until scrub grew. Kestrels can still be seen virtually everywhere in Britain and Ireland, but it has been overtaken by the buzzard and sparrowhawk as the commonest bird of prey, having declined by almost half since 1970, possibly because of a further loss of rough pasture and lack of voles through more deadly poisons. *Springwatch* has shown that mature songbirds and the chicks of larger birds, such as pheasants, as well as rodents, also form the kestrel's regular diet. In Wales and Northumberland's Kielder Forest, they themselves are preyed on by goshawks.

But for how much longer?

***

As will be clear, the two ornithologists who have most influenced these short essays have been 'BB' and Henry Douglas-Home. They never met, but their romantic attitudes and love of beauty were the same. 'BB' in a letter to David and Duncan McEwen when schoolboys wrote:

The lovely years lie in front of you. Remember above all things, that it is in nature and the natural world that the true joy of life can be experienced. My father once found the following words on a Cumberland gravestone and I have always prefaced my books with them:

The wonder of the world, the beauty and the
power, the shapes of things, their colours, lights,
and shades; these I saw. Look ye also while life lasts.

Derived from ROBERT BROWNING, 'Fra Lippo Lippi'

And Henry, to end *The Birdman*, wrote:

The world of birds may not be as efficient as the technical world of man, but it is an older, more lovely and, I think, a more peaceful place.

ACKNOWLEDGEMENTS

First to Richard Ingrams, who commissioned and named *The Oldie* Bird of the Month page on which this book is based, to Naim Attallah, first publisher of the magazine, and to Annie Soudain, first illustrator of Bird of the Month. To them, and *Oldie* publisher James Pembroke, Carry Akroyd and I owe the foundations and to James for smoothing the book's passage. For the support of succeeding *Oldie* editors, Alexander Chancellor, whose death is always lamented, and Harry Mount, who also wrote the generous introduction. And to the *Oldie* team, past and present: John Bowling, Liz Anderson, Jeremy Lewis, Deborah Maby, Lisa Martin, Ferdie Rous, Annabel Sampson, Nigel Summerley. For technological guidance, Gordon Monk. For reading the typescript, Alister Warman. For photograph, Richard Goodman. For securing Kathleen Jamie's permission, David McEwen. To Christian McEwen for introduction to the poetry of Gerry Cambridge and Thomas A. Clark. For greatly appreciated personal permissions: Gerry Cambridge (from *Aves*, 'Cinclus cinclus', 'Ardea cinerea', 'Haematopus ostralegus' (extract), 'Troglodytes troglodytes'; from *Notes for Lighting a Fire*, 'Blowing out an Egg'), Jane Clark (*The Alan Clark Diaries*, Kenneth Clark, *Another Part of the Wood, A Self Portrait*) Mark Cocker (*Birds Britannica*), Mary Colwell (*Curlew Moon*), Ned Denny (*B (after Dante)*), Peregrine Douglas-Home (Henry Douglas-Home, *The Birdman*), Sholto Douglas-Home (Robin Douglas-Home, *Hot for Certainties*), Max Egremont (John Wyndham, *Wyndham and Children First*), Rupert Lycett Green (John Betjeman, 'Death of King George V'), David Home (Lord Home, *Border Reflections*), Anne Humphreys (Bill Humphreys, 'Stormcock'), Helena McEwen (*The Big House*), Stephen Moss and Brent Westwood (*Tweet of the Day*), Adam Nicolson (*The Seabird's Cry*), Rupert Sheldrake (*Dogs That Know When Their Owners Are Coming Home*), Isabella Tree (*Wilding*). For personal contributions to the text: Wendy Baron, Rev P. Barron, Michael Chaplin, Paul Clements, Simon Courtauld, Caroline Douglas-Home, Merlin Errol, Joe Hardwicke, Alexander Hesketh, Freddy Hesketh, Paul Langley, James Le Fanu, John R. H. McEwen, Peregrine Moncreiffe, Robin Musson, Lawrence Preece, Matt Ridley, Richard Roberts, Roc Sandford, Bill Sylvester, Alister Warman, David Waters (Great Bustard Group), Tizzie Wells. For insights, contacts: Alexandra Connell, Duncan Davidson, Penelope Enthoven, Fiona Fraser, Leonie Gibbs, Ruth Guilding, Jane Haddington, Christabel Holland, Henry Keswick, Tessa Keswick, Brigid McEwen, Duncan McEwen, John Mclean, Jonny Minter, Miff Minter, Harry Mount, Julia Mount, Robin Page, James Pembroke, Kate Reid, Jason Russell, Talli and Isa Schoenborn-Burchheim, Paul Stancliffe (British Trust for Ornithology), John Stefanidis. Betty Stokes, Andrew Wilson, Badger Walker. For promotion: Chris Beetles, Amabel Lindsay. To Bryan Holden, whose anthologies and publication of the BB Society's *Sky Gipsy* magazine have done so much to protect and promote BB's legacy. For the authors of the contemporary books cited in the bibliography, above all to this book's progenitor, Mark Cocker's *Birds Britannica*. To Susie Chancellor for the ornithologist Robin Chancellor's set of *The Handbook of British Birds* and John Chancellor's biography of Audubon. To the staff of The London Library. To Clare Asquith and Sam Leith at *The Spectator*. To Mark Hedges, Mary Miers, Huon Mallalieu, Melanie Bryan and the team at *Country Life*, a mine of ornithological information, and to its principal bird writers David Profumo and Ian Morton. Climactically, to all those at Bloomsbury who made this book. Robin Baird-Smith for setting the ball rolling. Jim Martin for taking it on board and saying he wanted it to be 'glorious'. Claire Weatherhead for permissions advice and to Julie Bailey, who far surpassed her promise to be a 'protective' editor. For any oversight, we trust you know the authors well enough not to mind.

Bloomsbury Publishing would like to thank the following for providing permission to reproduce copyright material within this book:

'Woodpecker' from *Collected Poems* by Gerald Bullett is reprinted on page 7 by permission of Peters Fraser & Dunlop (www.petersfraserdunlop.com) on behalf of the Estate of Gerald Bulllett.

'The Dipper' from *Selected Poems* by Kathleen Jamie on page 11 is reproduced with the author's kind permission and by permission of the Licensor through PLSclear.

The extract on page 15 from *Landfill* by Tim Dee (Little Toller Books, 2018) is reproduced with kind permission of Tim Dee and Little Toller Books.

The extracts on pages 15, 19, 43, 65, 87, 95, 109 and 111 from *The Poems of Norman MacCaig*, edited by Ewen McCaig (Polygon, 2010), are reproduced by kind permission of Birlinn Ltd.

The extract from *The Rich: From Slaves to Super-Yachts* by John Kampfner on page 53 (Abacus, 2015) is reproduced by kind permission of David Higham Associates.

Extracts from *BB's Birds* (Roseworld Productions, 2007) on page 83, *Fisherman's Folly* (The Boydell Press, 1987) on page 93, *Sky Gipsy* (Eyre & Spottiswoode, 1942) on page 131, and *Tide's Ending* (Hollis & Carter, 1950) on pages 83, 117 and 131 are reproduced by kind permission of David Higham Associates.

Quotes from *The Peregrine* by J. A. Baker on pages 97, 123 and 129 are copyright © 1967 by J. A. Baker and are reprinted by kind permission of HarperCollins Publishers Ltd.

The extract from *Hellfire and Herring* on page 95 is copyright © Christopher Rush, 2006, and is reproduced by kind permission of Profile Books.

The extract from Peter Scupham's poem 'Kingfisher' on page 99 from his *Collected Poems*, published by Carcanet Press Limited, copyright © Peter Scupham 2002, is reproduced by kind permission of Carcanet Press Limited.

The poem from *Tormentil and Bleached Bones* by Thomas A. Clark on page 107 (Polygon, 2001) is reproduced by kind permission of Birlinn Ltd.

The text extracts from *The Gamekeeper* on page 119 are copyright © Portia Simpson 2017 and are reproduced by kind permission of Simon & Schuster UK Ltd.

***

Bloomsbury Publishing would also like to thank the following individuals for personally providing permission to include the extracts detailed below:

Mark Cocker for the quotes from his book with Richard Mabey *Birds Britannica* (Chatto & Windus, 2005) on pages 9, 31, 83, 103, 113, 125 and 133.

Gerry Cambridge for his poems 'Cinclus cinclus' on page 11, 'Ardea cinerea' on page 57, 'Troglodytes troglodytes' on page 103, and for the extract from 'Haematopus ostralegus' on page 117, all from *Aves* (Essence Press, 2007), and for his poem 'Blowing out an Egg' from *Notes for Lighting a Fire* (HappenStance Press, 2012) on page 25.

Adam Nicolson for the quote from his book *The Seabird's Cry* (William Collins, 2018) on page 13.

Rupert Sheldrake for the quote from his book *Dogs That Know When Their Owners are Coming Home: The Unexplained Powers of Animals* (Arrow, 2000) on page 17.

Ned Denny for the quote from *B (after Dante)* (Carcanet Press, 2020) on page 17.

Anne Humphreys for the extract from 'Stormcock' from *A Distant Cuckoo Calls* by Bill Humphreys (The Dovecote Press, 2008) on page 27.

Jane Clark for the quote from *Another Part of the Wood* by Kenneth Clark (Hamish Hamilton, 1995) on page 29 and for the quotes from *The Alan Clark Diaries* by Alan Clark (Weidenfeld & Nicolson, 1993) on pages 49 and 57.

Lord Home for the quote from *Border Reflections* (Collins, 1979) by his father Lord Home on page 29.

Sholto Douglas-Home for the quote from *Hot for Certainties* by Robin Douglas-Home (Pan, 1964) on page 61.

Peregrine Douglas-Home for the quotes from *The Birdman: Memories of Birds* (HarperCollins, 1977) by Henry Douglas-Home on pages 61 and 137.

Mary Colwell for the quote from her book *Curlew Moon* (William Collins, 2018) on page 65.

Max Egremont for the quote from *Wyndham and Children* First by John Wyndham (Macmillan, 1968) on page 71.

Isabella Tree for the quote from her book *Wilding* (Picador, 2018) on page 79.

Stephen Moss and Brett Westwood for the quote from their book *Tweet of the Day* (John Murray, 2016) on page 121.

Rupert Lycett Green for the quote from John Betjeman's poem 'Death of King George V' from *Collected Poems* (John Murray, 1948) on page 125.

Helena McEwen for the quote from her novel *The Big House* (Bloomsbury Publishing, 1999) on page 131.

BIBLIOGRAPHY

Ackermann, Jennifer, *The Genius of Birds*, Corsair, London, 2016

Aesop, *Fables*, Penguin Classics, London, 1996

Allen, Michael and Patel Ellis, Sonya, illustrations by Carry Akroyd, *Nature Tales, Encounters with Britain's Wildlife*, Elliott & Thompson, London, 2011

Armitage, Simon and Dee, Tim, *The Poetry of Birds*, Penguin Books, London, 2011

Baker, J. A., *The Peregrine*, Penguin Books, London, 1984

'BB' (Denys Watkins-Pitchford), *The Countryman's Bedside Book*, Eyre & Spottiswoode, London, 1941

—*Sky Gipsy*, Eyre & Spottiswoode, London, 1942

—*Brendon Chase*, Hollis & Carter, London, 1944

—*Tide's Ending*, Hollis & Carter, London, 1950

—*Dark Estuary*, Hollis & Carter, London, 1954

—*Little Grey Men*, Eyre & Spottiswoode, London, 1957

—*A Carp Water*, Putnam, London, 1958

—*The Autumn Road to the Isles*, Nicholas Kaye, London, 1959

—*The White Road Westwards*, Nicholas Kaye, London, 1961

—*September Road to Caithness and the Western Sea*, Nicholas Kaye, London 1962

—*The Naturalist's Bedside Book*, Michael Joseph, London, 1980

—*The Shooting Man's Bedside Book*, The Boydell Press, Woodbridge, 1986

—*Fisherman's Folly*, The Boydell Press, Woodbridge, 1987

Barham, Rev Richard Harris, *The Ingoldsby Legends*, Gresham, London, 1910

Beaton, Mrs Isabella, *The Book of Household Management*, S. O. Beeton, London, 1861

Benson, S. Vere, *The Observer's Book of Birds*, Frederick Warne, London, 1948

Bernard, Bruce (ed.), *Vincent by himself*, Orbis, London, 1985

Betjeman, John, *Collected Poems*, John Murray, London, 1948

Birkhead, Tim, *The Most Perfect Thing: Inside (and Outside) a Bird's Egg*, Bloomsbury Publishing, London, 2016

—*The Wisdom of Birds*, Bloomsbury Publishing, 2010

—*Bird Sense*, Bloomsbury Publishing, London, 2013

Blake, William, *Poetry & Prose*, The Nonesuch Press, London, 1927

Browne, Sir Thomas, *Notes and Letters of the Natural History of Norfolk*, Jarrold & Sons, London, 1902

Bronte, Emily, *Wuthering Heights*, Oxford University Press (OUP), Oxford, 1930

Buchan, John, *John MacNab*, Thomas Nelson and Sons, Edinburgh/London, 1945

Bullett, Gerald, *Collected Poems*, J. M. Dent & Sons, London, 1959

Burke, Janine, *Nest: The Art of Birds*, Allen & Unwin, London, 2012

Burns, Robert, *The Poems and Songs of Robert Burns*, OUP, Oxford, 1969

Cambridge, Gerry, *Aves*, Essence Press, Edinburgh, 2007

— *Notes for Lighting a Fire*, HappenStance Press, Glenrothes, 2012

Carey, John, *The Faber Book of Science*, Faber & Faber, London, 1995

Carr, Samuel, *The Poetry of Birds*, Batsford, London, 1976

Chapman, Abell, *Bird Life of the Scottish Borders*, Spredden Press, Stocksfield, 1990

Clare, John, *Bird Poems*, Folio Society, London, 2002

Clark, Alan, *The Alan Clark Diaries*, Weidenfeld & Nicolson, London, 1993

Clark, Kenneth, *Another Part of the Wood*, Hamish Hamilton, London, 1995

Clark, Thomas A., *Tormentil and Bleached Bones*, Polygon, Edinburgh, 1993

Cocker, Mark and Mabey, Richard, *Birds Britannica*, Chatto & Windus, London, 2005

Cocker, Mark and David Tipling (photographs), *Birds and People*, Jonathan Cape, London 2013

Colwell, Mary, *Curlew Moon*, William Collins, London, 2018

Cusa, Noel, *Tunnicliffe's Birds*, Victor Gollancz, London, 1984

Dante, *La Divina Commedia*, *Inferno*, canto v, *op cit* below

Darlington, Miriam, *Owl Sense*, Faber & Faber, London, 2018

Davies, Nick, *Cuckoo: Cheating by Nature*, Bloomsbury Publishing, London, 2015

Davies, W. H., *The Complete Poems of W. H. Davies*, Jonathan Cape, London, 1963

Dee, Tim, *Landfill*, Little Toller Books, Toller Fratrum, 2018

Denny, Ned, *B (after Dante)*, Carcanet Press, Manchester, 2020

Douglas-Home, Henry, *The Birdman*, Collins, London 1977

Douglas-Home, Robin, *Hot for Certainties*, Longmans, London, 1964

Egerton, Judy, *Turner: The Fighting Temeraire*, National Gallery Publications, London, 1995

Fisher, James, *Bird Recognition* (3 volumes), Penguin Books, Harmondsworth, 1947–55

Gibson, Graeme, *The Bedside Book of Birds*, Bloomsbury, London, 2005

Graves, Robert, *The White Goddess: A Historical Grammar of Poetic Myth*, Faber & Faber, London, 1948

Greenock, Francesca, *British Birds: their names, folklore and literature*, Christopher Helm, London, 1997

Grey of Fallodon, *The Charm of Birds*, Hodder & Stoughton, London, 1947

Hammond, Nicholas and Everett, Michael, *Birds of Britain and Europe*, Pan Books, London, 1980

Hardy, Thomas, *Collected Poems of Thomas Hardy*, Wordsworth Editions, Ware, 1994

Haupt, Lyanda Lynn, *Mozart's Starling*, Corsair, London, 2017

Hayward, John (ed.), *The Penguin Book of English Verse*, Penguin Books, London, 1958

Holden, Bryan, *The Art of BB*, Roseworld Productions, Solihull, 2016

—*BB's Birds*, Roseworld Productions, Solihull, 2007'

—*BB: A Symposium*, Roseworld Productions, Solihull, 2010

—*Confessions of a Coastal Gunner*, Roseworld Productions, Solihull, 2011

Home, Lord, *Border Reflections*, Collins, London, 1979

Hopkins, Gerard Manley, *The Poems of G. M. Hopkins*, OUP, Oxford, 1970

Hosking, Eric, *An Eye for a Bird*, Hutchinson, London, 1970

Hudson, W. H., *British Birds*, Longmans, Green, London, 1937

—*Birds and Green Places*, The Royal Society for the Protection of Birds, London, 1964

Humphreys, Bill, *A Distant Cuckoo Calls*, The Dovecote Press, Wimborne, 2008

Huxley, Elspeth, *Peter Scott*, Faber & Faber, London, 1993

Jacob, Violet, *The Scottish Poems of Violet Jacob*, Oliver and Boyd, Edinburgh, 1944

Jamie, Kathleen, *Selected Poems*, Picador Poetry, London, 2018

Lack, David, *The Life of the Robin*, Pallas Athene, London, 2016

Lawrence, D. H., *The Collected Poems of D. H. Lawrence* (2 volumes), William Heinemann, London, 1964

Lear, Edward, *A Book of Nonsense*, Littlehampton Book Services, Littlehampton, 1998

Lockley, Ronald, *The Island*, Penguin Books, London, 1969

—*Birds of the Sea*, Penguin, London, 1947

Lewis-Stempel, John, *Where Poppies Blow: The British Soldier, Nature, The Great War*, Weidenfeld & Nicolson, London, 2016

Lindley, Sir Francis, *A Diplomat Off Duty*, Ernest Been, London, 1947

Lockwood, W. B., *The Oxford Dictionary of British Bird Names*, OUP, Oxford, 1984

BIBLIOGRAPHY

McAdam Jr, E. L. and Milne, George, *Johnson's Dictionary, A Modern Selection*, Victor Gollancz, London, 1963

MacCaig, Norman, *The Poems of Norman MacCaig*, Polygon, Edinburgh, 2011

Macdonald Lockhart, James, *A Journey Through Birds*, Fourth Estate, London, 2016

McClatchy, Alfred, *On Wings of Song, Poems about Birds*, Alfred A. Knopf, New York, 2000

McEwen, Helena, *The Big House*, Bloomsbury, London, 2000

McEwen, John, *Bellany*, Mainstream Publishing, Edinburgh, 2013

McEwen, John H. F., *There is a Valley*, C.J. Cousland and Sons Ltd., Edinburgh, 1950

MacNeice, Louis, *Selected Poems of Louis MacNeice*, Faber & Faber, London, 1964

Meulen, Marcus van der, *The Brass Eagle Lectern in England,* Amberley Publishing, Stroud, 2017

Moss, Stephen, *A Bird in the Bush, A Social History of Bird Watching*, Aurum Press, London, 2004

Muirhead, George, *The Birds of Berwickshire* (2 volumes), David Douglas, Edinburgh, 1895

Munsterberg, Peggy, *The Penguin Book of Poetry*, Allen Lane, London, 1980

Murray, James A. H., *A New English Dictionary on Historical Principles* (21 volumes), Oxford at the Clarendon Press, Oxford, 1897

Mynott, Jeremy, *Birdscapes*, Princeton University Press, Woodstock, 2009

—*Birds in the Ancient World*, OUP, Oxford, 2018

Niccolai, Giulia, *Il Grande Angolo*, Feltrinelli, Milan, 1966

Nickerson, Sir Joseph, *A Shooting Man's Creed*, Swan Hill Press, Shrewsbury, 1989

Nicolson, Adam, *The Seabird's Cry*, William Collins, London, 2017

Poole, Charles Henry, *A treasury of bird poetry*. Simpkin, Marshall, Hamilton, Kent, London, 1923

Pope, Alexander, *The Poems of Alexander Pope*, Methuen, London, 1963

Potter, Stephen and Sargent, Laurens, *Pedigree: Words from Nature*, Collins, London, 1973

Powis, Llewelyn, *Earth Memories*, Redcliffe, Bristol, 1934

Preston, Alex and Gower, Neil, *As Kingfishers Catch Fire*, Corsair, London, 2017

Quinn, Tom, *BB: A Celebration*, Wharncliffe Publishing, Barnsley, 1993

Quiller-Couch, Sir Arthur, *The Oxford Book of English Verse 1250–1918*, Oxford at the Clarendon Press, Oxford, 1953

Reid-Henry, David, Harrison, Colin, *The History of the Birds of Britain*, Collins, London, 1988

Richardson, Rosamond. Illustrations by Carry Akroyd, *Waiting for the Albino Dunnock*, Weidenfeld & Nicolson, London, 2017

Rothenberg, David, *Why Birds Sing*, Allen Lane, London, 2005

Rothschild, Miriam, Clay, Theresa, *Fleas, Flukes and Cuckoos*, Macmillan, New York, 1957

Ruffer, Jonathan, *The Big Shots*, Quiller Press, London, 1977

Rush, Christopher, *Hellfire and Herring*, Profile Books, London, 2008

St John, Charles, *Wild Sports & Natural History of the Highland*s, T. N. Foulis, London & Edinburgh, 1919

Scott, Peter, *Morning Flight*, Country Life, London, 1944

Scott, Walter, *The Journal of Sir Walter Scott*, OUP, Oxford, 1972

Self, Andrew, *The Birds of London*, Bloomsbury, London, 2014

Shakespeare, William, *Tragedies, Comedies, Histories & Poems* (3 volumes), OUP, Oxford, 1962

Sheldrake, Rupert, *Dogs That Know When Their Owners Are Coming Home*, Arrow Books, London, 2011

Shute, Joe, *A Shadow Above: The Rise and Fall of the Raven*, Bloomsbury, London, 2018

Simpson, Portia, *The Gamekeeper*, Simon & Schuster, London, 2017

Skaife, Christopher, *The Ravenmaster*, 4th Estate, London, 2018

Smyth, Richard, *A Sweet Wild Note*, Elliott & Thompson, London, 2017

Stallard, Simon, *The Hidden Hut*, Harper Collins, London, 2018

Stevenson, Robert Louis, *The Collected Poems of Robert Louis Stevenson*, Edinburgh University Press, Edinburgh, 2003,

Strong, L. A. G., *The Brothers*, Victor Gollancz, London, 1932

Svenson, Lars, Mullarney, Killian & Zetterstrom, Dan, *Collins Bird Guide*, HarperCollins, London, 2009

Tennyson, Alfred, *The Works of Alfred Tennyson* (6 volumes), Strahan, London, 1873

The Holy Bible (King James version), Eyre & Spottiswoode, London, 1962

Thomas, Edward, *Collected Poems*, Faber & Faber, London, 1978

Thorburn, Archibald, *British Birds* (4 volumes), Longmans, Green, London, 1925

Toms, Mike, *Owls*, HarperCollins, London, 2014

Tree, Isabella, *Wilding*, Picador, London, 2018

Tudge, Colin, *The Secret Life of Birds*, Penguin, London, 2009

Tunnicliffe, C. F., *Shorelands Summer*, Orbis, London, 1985

—*Shorelands Winter Diary*, Robinson Publishing, London, 1992

Vaughan, Richard, *Plovers*, Terence Dalton, Lavenham, 1980

Westwood, Brett and Moss, Stephen, illustrations by Carry Akroyd, *Tweet of the Day*, Saltyard Books, London, 2014

Walton, Izaak, *The Compleat Angler,* J.M. Dent & Sons, London, 1970

Waters, Estlin and David, *The Great Bustard*, The Great Bustard Group, Salisbury, 2006

Watkins, Watkin, *The Birds of Tennyson*, R.H. Porter, London, 1903

White, Rev Gilbert, *The Natural History of Selborne*, Bell and Daldy, London, 1867

Witherby H. F., Jourdan F.C.R., Ticehurst N. F., Tucker B. W., *The Handbook of British Bird*s (5 volumes), H. F. & G. Witherby, London, 1938-55

Wodehouse, P. G., *Over Seventy*, Everyman, London, 2014

Wyndham, John, *Wyndham and Children First*, Macmillan, London, 1968

Vesey-Fitzgerald, Brian, *British Game*, Collins, London, 1946

Yeats, W. B, *Collected Poems of W. B. Yeats*, Vintage, London, 1992

Young, Andrew, *Collected Poems*, Rupert Hart-Davis, London, 1960

INDEX

INDEX

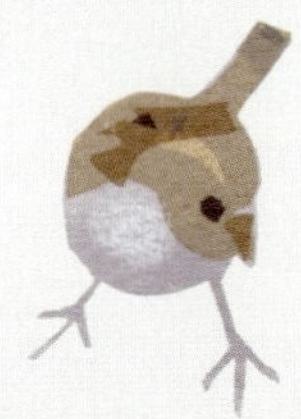